Student Course Guide
for

EXPLORING
Society

Introduction to Sociology

Third Edition

JANE A. PENNEY
Professor of Sociology
Dallas County Community College District

Dallas TeleLearning
R. Jan LeCroy Center for Educational Telecommunications
Dallas County Community College District

For use with the eighth edition of *Sociology in a Changing World* by
William Kornblum

THOMSON
WADSWORTH

Australia • Brazil • Canada • Mexico • Singapore • Spain • United Kingdom • United States

The Design and Production Team

Content Specialist:	Jane A. Penney, Professor of Sociology
Instructional Designer:	Suzanne Dunn, Ph.D.
Director of Production:	Craig Mayes
Producer/Director:	Ken Harrison
Production Coordinator:	Stephanie Bundschuh
Telecommunications Information Specialist:	Evelyn J. Wong
Research Assistant:	Laura Bohlcke

R. Jan LeCroy Center for Educational Telecommunications

President and Assistant Chancellor:	Pamela K. Quinn
Vice-President of Instruction:	James P. Picquet, Ph.D.
Dean of Distance Learning:	Edward C. Bowen
Dean of Business Affairs:	Dorothy J. Clark
Dean, Marketing and Community Relations:	Rachelle Howell
Director of Product Development:	Bob Crook

National Advisory Committee

New York City Technical College:	Victor Ayala
Middle Tennessee State University:	Carole Carroll
Rock Valley College:	Jerome Crane
Rose State College:	Terry Dean
City University of New York:	William Kornblum
Collin County Community College:	Tracey McKenzie Elliott
Portland Community College:	Rowan Wolf

DCCCD Advisory Committee

Richland College:	Kay Coder
El Centro College:	Glenn Currier
North Lake College:	Paul Magee
Cedar Valley College:	Tim Sullivan

Student Course Guide ISBN (10): 0-495-10098-6; ISBN (13): 978-0-495-10098-0
Copyright © 2008 by Thomson Publishing

Printed in the United States of America
10 9 8 7 6 5 4 3 2 1

THOMSON
WADSWORTH

Australia • Brazil • Canada • Mexico • Singapore • Spain • United Kingdom • United States

Contents

To the Student

Education is not the filling of a pail, but the lighting of a fire.
—William Butler Yeats

The ultimate goal of the educational system is to shift to the individual the burden of pursuing his education.
—John W. Gardner

Dear Student:

Your education represents an investment—not only in yourself, but also in your future. I hope this course, *Exploring Society, Introduction to Sociology*, helps you achieve your educational and personal goals!

Your exploration of sociology will take you on a fascinating journey. Sociology is an exciting discipline that helps us develop what sociologists call a *sociological imagination*. This enables us to explore our personal experiences within the context of our ever-shrinking world. Through your reading and viewing, you'll discover how the sociologist examines everyday activities and what makes sociology so relevant to our world.

One of the exceptional features of this course is that your journey into sociology is guided by sociologists from all over the United States, representing a rich diversity of ideas, heritages, genders, and ages! The contributions from these sociologists are simply outstanding. Meeting and interviewing these enthusiastic and dedicated colleagues renews my own sense of commitment to the discipline and to education.

But you'll also meet many people who are simply living their lives—some in extraordinary ways, some in very ordinary ways. These people contribute to the course by allowing their stories to be told. They help us understand how meaningful sociology is today and how important it is for our future!

As you embark on this exploration and journey into sociology, I hope you will ask yourself, "How does this apply to my life?" One of the challenges of education is to build that bridge from the concepts and theories of a discipline to the real lives of students. Beginning today, ask yourself how sociology is relevant in *your* life. Use the tools of sociology to appreciate your life and to celebrate the diversity of our global community.

Enjoy your exploration of society and journey into sociology!

—Jane A. Penney

About the Author:

Professor Penney is an educator who has spent her professional career in the classroom. She has a specialized interest in instructional design and learning styles. The heart of her approach in the classroom involves active learning methods. Jane believes the primary challenge for education is to make content real for students. She thinks of herself as a part of a learning community where both students and educators share in the teaching process.

Jane has been involved in distance learning since the inception of the R. Jan LeCroy Center for Educational Telecommunications. Currently, in addition to teaching in the classroom, she teaches both a video-based and online sociology course.

A Final Note:

With careful and thoughtful application of your time and energy to the material presented in this course, you should have a rewarding experience in the broadest sense of that term. I, along with other members of the production team, have put forth our best efforts to create a quality course. However, my experience teaches me that any course can be improved, so I encourage you to share any ideas about it with me. Please send your comments to Jane A. Penney, R. Jan LeCroy Center for Educational Telecommunications, 9596 Walnut Street, Dallas, TX 75243-2112, or email me at jpenney@dcccd.edu.

Student Course Organization

Exploring Society, Introduction to Sociology is designed as a comprehensive learning package consisting of four elements: student course guide, textbook, video programs, and interactive activities.

STUDENT COURSE GUIDE

The student course guide for this course is:

Penney, Jane A. *Student Course Guide for Exploring Society, Introduction to Sociology*. 3rd ed. Belmont, CA: Wadsworth Publishing, 2008. ISBN (10): 0-495-10098-6; ISBN (13): 978-0-495-10098-0

The student course guide serves as your daily instructor. If you follow the suggested Study Guidelines carefully, you should successfully accomplish all the requirements for this course. (See the section entitled "Student Course Guidelines.)

TEXTBOOK

In addition to the student course guide, the required book for this course is:

Kornblum, William. *Sociology in a Changing World*, 8th ed. Belmont, CA: Wadsworth Publishing, 2008. ISBN (10): 0-495-09635-0; ISBN (13): 978-0-495-09635-1

This comprehensive, student-friendly introductory textbook emphasizes the reality of social change and its impact on individuals, groups, and societies throughout the world. The text carefully balances contemporary and classical theory and research, with special attention to the contributions of female and minority social scientists and cross-cultural studies.

VIDEO PROGRAMS

The video program series for this course is:

Exploring Society, Introduction to Sociology

Each video program has a corresponding student course guide assignment that includes text readings associated with the lesson topic. The video programs are presented in a documentary format and are designed to engage the viewer through analysis and perspectives of the issues being discussed. Watch them closely.

If the video programs are broadcast more than once in your area, or if DVDs, CDs, videotapes, streaming video, or audiotapes are available at your college, you might find it helpful to watch the video programs more than once or listen to an audio tape for review. Also, you may tape the programs for review or viewing at a more convenient time. Since examination questions will be taken from the video programs as well as from the reading, careful attention to both is vital to your success.

COMPUTER-BASED ACTIVITIES

Self-graded interactive exercises, pre- and post-self-assessments, and case-based, problem-solving scenarios are available to students whose institutions have opted to offer these. These activities are useful for reinforcement and review of lesson content and learning objectives. The interactive activities are offered in two formats: CD-ROM and Internet. Ask your instructor how to access these activities if they are listed in your syllabus as a course requirement.

Student Course Guidelines

Follow these guidelines as you study the material presented in each lesson:

1. OVERVIEW—
   Read the Overview for an introduction to the lesson material.

2. LESSON ASSIGNMENT—
   Review the Lesson Assignment in order to schedule your time appropriately. Pay careful attention—the titles and numbers of the textbook chapters, the student course guide lessons, and the video programs may be different from one another.

3. LESSON GOAL—
   Review the Lesson Goal to learn what you are expected to know or be able to do upon completion of the lesson.

4. LESSON LEARNING OBJECTIVES—
   Review the Learning Objectives to guide you in successfully mastering the lesson content and achieving the Lesson Goal.

5. REVIEW—
   The following steps are intended to help you learn the material in this lesson. To maximize your learning experience:
   a. Scan the Lesson Focus Point questions.
   b. Read the assigned text pages.
   c. View the video.
   d. Write answers to the Lesson Focus Point questions.
   e. Complete the Related Activities assigned by your instructor. If none are assigned, use them to help you review the lesson material.
   f. Take the Practice Test and check your answers with the Answer Key located at the end of the lesson.

6. LESSON FOCUS POINTS—
   Pay attention to the Lesson Focus Points to get the most from your reading and viewing. You may want to write responses or notes to reinforce what you learn as you progress through the lesson material.

7. RELATED ACTIVITIES—
    The Related Activities are not required unless your instructor assigns them. They are offered as suggestions to help you learn more about the material presented in the lesson.

8. PRACTICE TEST—
    The Practice Test will help you evaluate your understanding of the material in this lesson. Use the Answer Key located at the end of the lesson to check your answers or reference material related to each question.

9. ANSWER KEY—
    The Answer Key provides answers and references for the Practice Test questions.

Lesson 1

Why Sociology?

OVERVIEW

Why study sociology? What will you gain by developing a *sociological imagination*? How does studying sociology help you to better understand how people interact socially within groups, organizations, and society, and why people make the choices they do? As you study sociology, you will develop your sociological imagination and gain a better understanding of your social world.

What attracts people to the discipline of sociology? What types of events and conditions do sociologists study? What makes the way that sociologists study people unique and different from the simple observations we make every day?

Sociology provides tools that can help us understand the forces at work in society. By using these tools, sociologists challenge us to explore our social world and the larger, global environment in which we live.

What is *visual sociology*? In the video lesson, you will meet sociologist Carol Chenault. She travels the world with her photographer husband and captures on film images of unique cultures. Chenault uses these photographs in her teaching to help her students gain a better understanding of distant cultures and peoples. What do such visual images teach us about cultures, people, and life? Are pictures truly worth "a thousand words"?

Also in this lesson, you will learn about the men and women who provided the foundation and contributed to sociology becoming a discipline that is recognized by cultures and societies throughout the world, like Harriet Martineau, a sociologist of the early nineteenth century. Today, there are many women who have spent their professional lives making important sociological contributions. Sociology has a history of development that is important because it helps us understand how it came to be the widely accepted discipline it is today.

LESSON ASSIGNMENTS

Text: William Kornblum: *Sociology in a Changing World*, Chapter 1, "Sociology: An Introduction," pp. 2–12 and pp. 19–21

Video: "Why Sociology?" from the series *Exploring Society, Introduction to Sociology*

LESSON GOAL

After completing this lesson, you will know how sociology developed as a discipline and have an appreciation for the sociological way of looking at things.

LESSON LEARNING OBJECTIVES

1. Identify sociology as a discipline.
2. Describe *visual* sociology.
3. Describe the development of sociology as a discipline.
4. Explain how a sociological imagination increases awareness of self and society.

REVIEW

The following steps are intended to help you learn the material in this lesson. To maximize your learning experience:

a. Scan the Lesson Focus Point questions.
b. Read the assigned text pages.
c. View the video.
d. Write answers to the Lesson Focus Point questions.
e. Complete the Related Activities assigned by your instructor. If none are assigned, use them to help you review the lesson material.
f. Take the Practice Test and check your answers with the Answer Key located at the end of the lesson.

LESSON FOCUS POINTS

1. Why is sociology a discipline?
2. How did sociology develop as a discipline?
3. How does sociology differ from common sense?
4. What topics do sociologists study?
5. What is *sociological imagination*?
6. How does sociology differ from other disciplines?
7. What are the three different levels of society's complexity that sociologists study? Be able to provide examples.

---

8. How does a systematic observation of society contribute to our understanding of society?
9. What is *visual sociology*?
10. How does visual sociology provide important clues about society?
11. How did the following people contribute to sociology?
    A. Auguste Comte
    B. Émile Durkheim
    C. Max Weber
    D. Karl Marx
    E. W.E.B. Du Bois
    F. Oscar Lewis
    G. Edward Franklin Frazier
    H. Jane Addams
    I. Ida B. Wells-Barnett

12. What is the importance of sociology in cultures other than American culture?
13. How are the careers of Harriet Martineau and Carol Chenault similar? How are they different?
14. What are the sociological implications of a parade?
15. What is the role of the sociologist in identifying his/her own values and prejudices?

RELATED ACTIVITIES

1. Attend a local, community parade. After viewing the parade with your sociological imagination, write a paper about what you discovered about your community.

2. Find a visual image (in a magazine or book) that you think reflects something important about society. Copy the picture and write a paper about what you have learned about society from visual images.

3. View a current, popular movie. Write about how the movie does or does not reflect society. Give specific examples from the movie.

4. Describe how gang behavior might be studied on the *micro, middle*, and *macro* levels of analysis in sociology. What are some of the questions that a sociologist would ask at each of those levels of analysis?

PRACTICE TEST

<u>Multiple Choice</u>
Choose the BEST answer.

1. Sociologists are concerned with how _____ influence our lives as individuals.
   A. social conditions
   B. environments
   C. psychological processes
   D. artifacts

2. Which of the following areas of inquiry would NOT be studied by sociologists?
   A. Behavior of work groups
   B. Changing relationships between men and women
   C. The activities of gangs in an urban area
   D. How regions of the brain function

3. Sociology involves the _____ of social behavior.
   A. scientific study
   B. individual application
   C. historical evolution
   D. individual selection

4. To study the seating pattern that exists in a family, a sociologist would study the patterns at the _____ level.
   A. macro
   B. middle
   C. micro
   D. informal

5. Information based on _____ is verifiable and testable.
   A. empirical evidence
   B. common sense
   C. informal observation
   D. scientific laboratories

6. _____ sociology involves observational techniques of photography and videotape.
   A. Popular
   B. Visual
   C. Cultural
   D. Reflective

7. Which of the following could NOT be conveyed as an element of visual sociology?
   A. Characteristics of culture
   B. The social structure of a community
   C. Patterns of inequality
   D. An interpretation of the image

8. Which of the following nineteenth-century sociologists would NOT be one most sociologists credit with being the most influential?
   A. Karl Marx
   B. Jane Addams
   C. Émile Durkheim
   D. Max Weber

9. Early sociological thinkers understood the importance of applying _____ to the study of society.
   A. numerical ideology
   B. psychological profiles
   C. informal observation
   D. the scientific method

10. The word "sociology" was coined by French philosopher, _____.
    A. Auguste Comte
    B. Comte August
    C. Simon Compte
    D. Renon Revoir

11. In the late nineteenth century in France, _____ wrestled with the issue of modernization.
    A. Max Weber
    B. Karl Marx
    C. Auguste Comte
    D. Émile Durkheim

12. _____, an African-American woman, wrote extensively about discrimination and racism and became a spokesperson for the civil rights of African-Americans and women.
    A. Harriet Martineau
    B. Carol Chenault
    C. Jane Addams
    D. Ida B. Wells-Barnett

13. _____ is believed to be the first woman to contribute to the emerging field of sociology.
    A. Jane Addams
    B. Harriet Martineau
    C. Carol Chenault
    D. Janet Gibson

14. In the 1920s, the _____ and _____ were the centers of American sociological research.
    A. University of Chicago; Columbia University
    B. University of New York; Cornell University
    C. University of Maine; Princeton
    D. College of Illinois; Yale

15. _____ was founded by Jane Addams to serve the poor immigrant neighborhood in Chicago.
    A. Hudson Bay Settlement
    B. Chicago Circle
    C. Hampton House
    D. Hull House

16. The field of human ecology was developed at the _____ by Robert Park and Ernest Burgess.
   A. University of Chicago
   B. University of Illinois
   C. Howard University
   D. South Carolina University

17. Often referred to as a sort of mother figure in the sociological tradition, _____ helped others understand that sociology had a purpose and could provide a set of explanations that other fields could not.
   A. Jane Addams
   B. Ida B. Wells-Barnett
   C. Jane B.Wells
   D. Ida Simpkins

18. In observing a Rose Bowl parade, a sociologist might observe _____.
   A. society's values of competition and success
   B. symbols of American culture
   C. issues related to gender inequality and social injustice
   D. all of the above

19. Which of he following is NOT accomplished after developing a sociological imagination?
   A. Your ability to participate in social life
   B. Your ability to step back from social life
   C. Your ability to analyze the broader meanings of social life
   D. Your ability to get along with other people

20. Developing and strengthening your sociological imagination means that you will _____.
   A. learn how to apply evidence to your views
   B. modify your views based on evidence
   C. both A and B
   D. none of the above

Essay Question

Answer the following question using complete sentences in a well-developed essay.

1. Discuss the benefits of developing a sociological imagination. Give an example of how using a sociological imagination is beneficial.

ANSWER KEY

The following provides the answers and references for the practice test questions.

Multiple Choice:
1. A............LO 1 ....textbook, p. 3
2. D ....LO 1 ....textbook, pp. 4–5
3. A ....LO 1 ....textbook, p. 4; video segment 2
4. C ....LO 1 ....textbook, pp. 5–6; video segment 2
5. A ....LO 1 ....textbook, p. 10; video segment 2
6. B............LO 2 ....textbook, p. 18; video segment 3
7. D ....LO 2 ....textbook, p. 18; video segment 3
8. B............LO 3 ....textbook, pp. 7–9
9. D ....LO 3 ....textbook, p. 7; video segment 4
10. A............LO 3 ....textbook, p. 7; video segment 4
11. D ....LO 3 ....textbook, p. 8; video segment 4
12. D ....LO 3 ....textbook, p. 9
13. B............LO 3 ....textbook, p. 9; video segment 5
14. A............LO 3 ....textbook, p. 10
15. D ....LO 3 .... video segment 4
16. A............LO 3 ....textbook, pp. 10–11; video segment 4
17. B............LO 3 .... video segment 4
18. D ....LO 4 ....textbook, pp. 2 and 18–19
19. D ....LO 4 ....textbook, pp. 3–4
20. B............LO 4 ....textbook, p. 10

Essay Question

Answers should include the following types of statements/points:

1. ....LO 4 ....textbook, p. 3–4
    Developing a sociological imagination enables a person to participate in social life and step back and analyze the broader meanings of what is going on. Example: a parade, while fun and entertaining, also presents insights into the society and community. How do the people related to one another? What stereotypes persist? How inclusive of diversity are the participants? Who sponsors the parade? Are there clues about gender roles?

LESSON CONTRIBUTORS

Victor A. Ayala, Professor of Sociology, New York City Technical College, Brooklyn, NY

Juan Battle, Professor of Sociology, Hunter College, New York, NY

Carol Deming Chenault, Instructor of Sociology, Calhoun Community College, Decatur, Alabama

Kay Coder, Professor of Sociology, Richland College, Dallas, TX

Troy Duster, Professor of Sociology, New York University, New York, NY

Susan Brown Eve, Professor of Applied Gerontology and Sociology, University of North Texas, Denton, TX

Paul Geisel, Professor of Urban Affairs, University of Texas at Arlington, Arlington, TX

Arlie Russell Hochschild, Professor of Sociology, University of California—Berkeley, Berkeley, CA

Leslie Irvine, Assistant Professor of Sociology, University of Colorado—Boulder, Boulder, CO

Ed Joubert, Retired Sociology Professor, University of Louisiana—Lafayette, Carencro, LA

Elaine Bell Kaplan, Associate Professor of Sociology, University of Southern California, Los Angeles, CA

William Kornblum, Professor of Sociology, City University of New York, New York, NY

Michele Lamont, Professor of Sociology, Princeton University, Princeton, NJ

Paul Magee, Professor of Sociology, North Lake College, Irving, TX

Tom Mayer, Professor of Sociology, University of Colorado—Boulder, Boulder, CO

Mary Patillo, Associate Professor of Sociology, Northwestern University, Evanston, IL

Judith Perrolle, Associate Professor of Sociology, Northeastern University, Boston, MA

Claire Renzetti, Professor and Chair of Sociology, St. Joseph's University, Philadelphia, PA

George Ritzer, Professor of Sociology, University of Maryland, College Park, MD

Rudy Ray Seward, Professor and Associate Chair of Sociology, University of North Texas, Denton, TX

Mary Virnoche, Assistant Professor, Humboldt State University, Denver, CO

Terry Williams, Associate Professor, New School University, New York, NY

Alford Young, Assistant Professor, University of Michigan, Ann Arbor, MI

Special acknowledgement and thanks go to:
   Lafayette Convention and Visitor's Commission, Lafayette, Louisiana
   Randol's Restaurant, Lafayette, Louisiana

Lesson 2

Sociological Perspectives

OVERVIEW

Examining our social world is not a haphazard and naïve activity. It is a disciplined and critical process. To explore the multidimensional nature of our society, sociologists use *sociological perspectives*. These perspectives provide valuable insights into the social structures which people create to maintain order in their relationships. Sociologists use the perspectives to explore human conflict associated with control and power and how we construct social meanings in our groups, communities, and society.

In this video lesson, you will learn about each of the sociological perspectives and their respective characteristics. The functionalist perspective stresses the structure and organization of our society and helps us understand how all the intricate workings of our society function to meet our needs. Order, structure, purpose—these are the key words associated with the functionalist perspective.

The conflict perspective focuses on social change. The key words are conflict, power, and control. When using the conflict perspective, a sociologist would ask questions like: Who has the power? Who is in control? Who has the most to gain? Often issues related to wealth and access to powerful resources provide the research questions most appropriate for the conflict perspective. Emerging from this perspective, the feminist perspective examines issues related to gender and social control, power, and access.

The interactionist perspective is about how social interactions and interplay define our social world. The key words are communication, interpretation, and shared meanings. When using the interactionist perspective, a sociologist would ask questions like: How do people communicate, negotiate, share, or compete? What is the result of the interactions among people? One area of study that is best probed using the interactionist perspective is the power of language and labels.

In this video lesson, you will see how using the sociological perspectives enhances our understanding of an artistic depiction of the social life on a sugar plantation in old Mexico. You will also learn how using the sociological perspectives broadens our interpretation of major news events, like the story of Elian Gonzalez. Beginning with his flight from Cuba, to his ultimate reunion with his father, and return to Cuba, each sociological perspective asks different sorts of questions and examines different levels of social interaction throughout the event

Whether it is paintings or news events we're trying to understand, the sociological perspectives can help us assemble the pieces of the multidimensional puzzle we call society.

LESSON ASSIGNMENTS

Text: William Kornblum: *Sociology in a Changing World*, Chapter 1, "Sociology: An Introduction," pp. 12–20

Video: "Sociological Perspectives" from the series *Exploring Society, Introduction to Sociology*

LESSON GOAL

After completing this lesson, you will be able to distinguish among the sociological perspectives in sociology.

LESSON LEARNING OBJECTIVES

1. Describe the major sociological perspectives.
2. Describe the feminist sociological perspective.
3. Interpret an event using the sociological perspectives.
4. Explain how the sociological perspectives contribute to a critical understanding of society.

REVIEW

The following steps are intended to help you learn the material in this lesson. To maximize your learning experience:

a. Scan the Lesson Focus Point questions.
b. Read the assigned text pages.
c. View the video.
d. Write answers to the Lesson Focus Point questions.
e. Complete the Related Activities assigned by your instructor. If none are assigned, use them to help you review the lesson material.
f. Take the Practice Test and check your answers with the Answer Key located at the end of the lesson.

LESSON FOCUS POINTS

1. What is *interactionism*?
2. What is *functionalism*?
3. What is the conflict perspective?
4. What is the feminist perspective?
5. How are the sociological perspectives alike and/or different?
6. How do sociologists interpret social events using the interactionist perspective? What questions might they ask? What elements of the events do they observe?
7. How do sociologists interpret social events using the functionalist perspective? What questions might they ask? What elements of the events do they observe?
8. How do sociologists interpret social events using the conflict perspective? What questions might they ask? What elements of the events do they observe?
9. Why is a critical understanding of social behavior important?
10. How do the three sociological perspectives contribute to our critical understanding of social relationships and groups?
11. What are the benefits of examining issues from more than one perspective?
12. What issues would be examined using the feminist perspective?

RELATED ACTIVITIES

1. With which of the sociological perspectives do you identify the most; that is, which is most similar to your view of society and social life? Explain what life experiences have influenced your thinking.

2. Look in the newspaper for an article about a current event. Interpret that event using the sociological perspectives. Point out the key questions/issues that each of the sociological perspectives would help sociologists explore.

3. Write an essay that supports the value of seeking a multidimensional view of society.

PRACTICE TEST

Multiple Choice
Choose the BEST answer.

1. Sociological _____ are sets of ideas and theories that sociologists use in attempting to understand various problems of human society.
   A. theories
   B. correlations
   C. perspectives
   D. tools

2. Which sociological perspective views social order and social change as resulting from repeated interactions between people?
   A. Functionalism
   B. Conflict perspective
   C. Feminist perspective
   D. Interactionism

3. At which level of analysis of social life does the interactionist perspective usually occur?
   A. Interpersonal relationships
   B. Bureaucratic interactions
   C. Governmental interactions
   D. All of the above

4. The _____ theory focuses on what people seem to be getting out of their interactions and what they are contributing.
   A. exchange
   B. rational
   C. deductive
   D. balance

5. _____ calls attention to how social life is constructed through mundane acts of social communication.
   A. Symbolic constructivism
   B. Symbolic interactionism
   C. Interactive symbolism
   D. Interpretive symbolic theory

6. The _____ perspective asks how society manages to carry out the jobs it must perform in order to maintain social order.
   A. conflict
   B. feminist
   C. interactionist
   D. functionalist

7. Which of the following would be a visual metaphor for the functionalist arrangement of modern work life?
   A. Working in a home office
   B. Working in cubicles in an open office space
   C. Working in individual, large offices in privacy
   D. All of the above

8. When various structures of society become poorly integrated because of social change, the formerly useful functions can become _____.
   A. mysfunctional
   B. dysfunctional
   C. segregated
   D. non-functional

9. In an agrarian society, a large family is considered valuable with many members contributing to the well-being of the family when lots of hands were needed to harvest crops. In an industrial society, a large family might be considered _____ .
   A. dysfunctional
   B. functionalistic
   C. interactive
   D. conflictism

10. Which of the following events of the 1960s provided impetus to the conflict perspective of sociology?
    A. The women's movement
    B. Anti-war demonstrations
    C. Environmental protests
    D. All of the above

11. The concept of _____ is central in conflict theory.
    A. power
    B. human rights
    C. equality
    D. authority

12. _____ was responsible for the foundation of ideas for the conflict perspective of sociology.
    A. Emile Durkheim
    B. Karl Marx
    C. Auguste Comte
    D. Georg Simmel

13. _____ is the ability of an individual or group to change the behavior of others.
    A. Authority
    B. Control
    C. Power
    D. None of the above

14. Which of the following issues would be examined using the feminist perspective?
    A. Gender and social control
    B. Gender and power
    C. Gender and access
    D. All of the above

15. The _____ perspective presents the idea that the sociological study of society has been the study of men in society, not the entire society.
    A. masculinist
    B. feminist
    C. individualist
    D. gender

16. When Karl Marx studied work, he completely ignored _____.
    A. factory work
    B. men's work
    C. blue collar work
    D. work in the home

17. One question explored by the feminist perspective is, _____?
    A. why do men make more money than women
    B. why hasn't a woman been president of the United States
    C. why isn't the United States Senate composed of more women
    D. all of the above

18. One of the challenges for sociologists is to attempt to _____.
    A. become involved in the event
    B. just deal with facts
    C. step back and look at a situation
    D. change society

19. Part of the reason the Elian Gonzales story became so emotionally charged is because _____.
    A. many in Miami's Cuban community arrived in the same manner
    B. people distrust the United States government
    C. Elian Gonzales had lost his mother
    D. Elian's father was unfit

20. For Cuban Americans, Elian Gonzales was a symbol of _____.
   A. anti-American sentiments
   B. the struggle against Fidel Castro
   C. the old Cuba they knew
   D. all of the above

Essay Question

Answer the following question using complete sentences in a well-developed essay.

1. Using the sociological perspectives to examine a situation provides one with a multidimensional understanding of that situation. What is the central focus of each sociological perspective? What are the advantages of using more than one sociological perspective?

ANSWER KEY

The following provides the answers and references for the practice test questions.

Multiple Choice:

1. C ....LO 1 ....textbook, pp. 12–13
2. D ....LO 1 ....textbook, pp. 13–14; video segment 2
3. A ....LO 1 ....textbook, p. 13; video segment 2
4. A ....LO 1 ....textbook, p. 15
5. B ....LO 1 ....textbook, p. 16
6. D ....LO 1 ....textbook, pp. 16–17; video segment 2
7. B ....LO 1 ....textbook, p. 17
8. B ....LO 1 ....textbook, p. 17
9. A ....LO 1 ....textbook, p. 17
10. D ....LO 1 ....textbook, p. 18
11. A ....LO 1 ....textbook, p. 19
12. B ....LO 1 ....textbook, p. 18
13. C ....LO 1 ....textbook, p. 19
14. D ....LO 2 .... student course guide lesson overview, p. 11
15. B ....LO 2 .... video segment 3
16. D ....LO 2 .... video segment 3
17. D ....LO 2 .... video segment 3
18. C ....LO 3 .... video segment 4
19. A ....LO 3 .... video segment 4
20. B ....LO 3 .... video segment 4

Essay Question

Answers should include the following types of statements/points:

1. ....LO 4...textbook, p. 12–20
   <u>Central focus— sociological perspectives</u>
   - Functionalist: looks at issues involved with order, structure; examines the function of society's organization and structures in resolving issues.
   - Conflict: looks at issues centered around power, control, and access; examines who has control and power.
     - ◆ Feminist: examines some issues as the conflict perspective but in relation to gender.
   - Interactionist: looks at shared meanings, communication and interpretations; examines how different groups can interpret the same event in very different terms.

   <u>Advantages:</u>
   - No one perspective can provide a complete explanation—a deeper understanding is gained by using all the sociological perspectives.
   - Provides insights into how different groups interpret an event very differently.
   - Our society is too complex for one perspective; a multidimensional view of society is desirable.
   - Helps you gain an understanding of diversity of views and attitudes.

LESSON CONTRIBUTORS

Wanda Alderman-Swain, Professor of Sociology, Howard University, Washington, DC

Guillermo J. Grenier, Associate Professor/Director, Florida International University, Miami, FL

Leslie Irvine, Assistant Professor of Sociology, University of Colorado—Boulder, Boulder, CO

Lisandro Pérez, Sociologist, Florida International University, Miami, FL

Claire M. Renzetti, Professor and Chair of Sociology, St. Joseph's University, Philadelphia, PA

Lesson 3

Sociological Inquiry

OVERVIEW

Everyday, we hear statistics about various products and events that are intended to influence our behavior. On television, a particular product is said to be 99% effective, or consumers select one product over a like product 80% of the time, or 84% of voters are projected as supporting a particular candidate. We are surrounded by such information and often assume such statistics to be the result of scientific investigation. But are they?

Often, we draw conclusions about our world by making casual observations. For example, we ask friends for their opinions; we base our assumptions on an article we read; we listen to newscasters and believe what they tell us. We prepare for a test by completing a practice test to determine how ready we are for the real test. If we do well on the practice test, we conclude that we are ready for the test; if we do poorly on the practice test, we draw a different conclusion. Are such conclusions accurate?

While such casual approaches seem helpful, there are more formal methods that sociologists use in research. Sociologists know that their studies are going to be scrutinized by other sociologists. Applying the philosophy of empiricism to the investigation of human group behavior is what distinguishes sociological scientific research from casual observation. The use of scientific research is part of what makes sociology a recognized discipline.

In this lesson, you will meet sociologists who are involved in very different research projects. Rob Gardner, a sociologist in Colorado, loves blue grass music and uses a qualitative research method to gather insights about this community of musicians. Juanita Firestone and Richard Harris use quantitative research to explore sexual harassment in the military. Sociologists William Kornblum and Terry Williams use a mixed methods approach, combining qualitative and quantitative methods to probe the world of the homeless in New York City.

LESSON ASSIGNMENTS

Text: William Kornblum: *Sociology in a Changing World*, Chapter 2, "The Tools of Sociology," pp. 23–42

Video: "Sociological Inquiry" from the series *Exploring Society, Introduction to Sociology*

LESSON GOAL

After completing this lesson, you will be able to explain how the process by which sociologists study group behavior differs from casual observations and common sense conclusions.

LESSON LEARNING OBJECTIVES

1. Discuss why empirical research is important to the discipline of sociology.
2. Using an example, explain the process of quantitative research.
3. Using an example, explain the process of qualitative research.
4. Discuss the mixed methods approach to sociological inquiry.

REVIEW

The following steps are intended to help you learn the material in this lesson. To maximize your learning experience:

  a. Scan the Lesson Focus Point questions.
  b. Read the assigned text pages.
  c. View the video.
  d. Write answers to the Lesson Focus Point questions.
  e. Complete the Related Activities assigned by your instructor. If none are assigned, use them to help you review the lesson material.
  f. Take the Practice Test and check your answers with the Answer Key located at the end of the lesson.

LESSON FOCUS POINTS

1. What are some limitations in accepting common sense conclusions as truths?
2. What is *empiricism*?
3. What is the *scientific method*?
4. What are the steps in the scientific method?
5. Why is empirical research important to sociology?
6. What is the difference between qualitative and quantitative research?
7. What is *ethnography*?
8. Following the steps of scientific inquiry, describe the research of Juanita Firestone and Richard Harris.
9. Following the steps of scientific inquiry, describe the research of Rob Gardner.
10. Following the steps of scientific inquiry, describe the research of William Kornblum and Terry Williams.
11. What is the mixed methods approach in sociological inquiry?
12. What are the advantages of each type of sociological inquiry?
13. What makes the information gained through sociological inquiry different from information you read in everyday newspapers and magazines?
14. Why was Emile Durkheim's study of suicide important?
15. What is the difference between a dependent and an independent variable?
16. What are the basic methods used to conducting research in sociology?
17. What is the difference between a control group and an experimental group in research?
18. What is the Hawthorne effect?
19. What are examples of survey research?
20. What are some ethical considerations to consider when conducting sociological research?
21. What are the definitions of privacy, confidentiality, and informed consent?
22. What is a theory?
23. Why is the U.S. Census important to sociologists?
24. What is visual sociology?
25. What research was conducted by Solon Asch in the early 1950s? What is the significance of his research?
26. What does the term *correlation* mean?
27. Why are mapping techniques important to sociologists?

RELATED ACTIVITIES

1. If you had a large grant to do a sociological study, what subject, issue, or problem would you be interested in studying? Why did you choose that topic; that is, what was there about your experiences, background, and interests that caused you to focus on that topic?

2. Your neighborhood offers many opportunities to do sociological research using unobtrusive measures. Describe at least two possible studies that could be done in your neighborhood using this method. How would you gather your evidence? What is it about people's lifestyles, groups, or social relationships that you could discover using this technique?

3. Design a questionnaire to discover student attitudes toward telecourses. Compose at least four questions to obtain the information you need. You may use "open questions" or "closed questions." Describe the kind of sample you would use, as well as how you would identify your respondents and acquire the responses to the questionnaire, that is, how you would gather your data.

4. After viewing the video program, describe your reaction to, and evaluation of, the research done by Professors William Kornblum and Terry Williams.

5. Select a research study reported in a sociological journal. Identify the parts of the research that correspond to the steps of the scientific method.

6. Select a topic that interests you. Read an article in a popular magazine about that topic. Then find and read a research article about the same topic in a sociological journal. Describe the differences you found between the two sources of information.

PRACTICE TEST

Multiple Choice
Choose the BEST answer.

1. Which of the following should you consider when formulating research questions?
   A. What has already been published
   B. Time and resources
   C. Your own interest
   D. All of the above

2. Which of the following statements is supported by research?
   A. Couples who live together before marriage have more successful marriages than those couples who do not.
   B. Our welfare system encourages teenage girls to have more children.
   C. The differences in the behavior of men and women is simply human nature.
   D. There is no significant statistical difference in the divorce rate of couples who live together before marriage and those who do not live together before marriage.

3. Sociologists use the _____ process to collect and analyze information.
   A. research
   B. survey
   C. social science
   D. sociology

4. The methodology that sociologists use to study issues means that the study can be _____ by other sociologists.
   A. observed
   B. written
   C. replicated
   D. collected

5. _____ sociological research relies on statistical analysis and its focus is usually narrow and concise.
   A. Qualitative
   B. Quantitative
   C. Sample
   D. Observation

6. To the sociologist, which research method makes important contributions to the discipline of sociology?
   A. Ethnography
   B. Qualitative research
   C. Quantiative research
   D. All of the above contribute

7. Which of the following statements does NOT represent a conclusion reached in Emile Durkheim's study of suicide?
   A. Suicide rates should be higher for people without children.
   B. Suicide rates should be higher for married people.
   C. Suicide rates should be higher for people with higher levels of education.
   D. Suicide rates should be higher in Protestants than in Catholic communities.

8. In Emile Durkheim's study of suicide, the suicide rate is the _____ variable.
   A. dependent
   B. independent
   C. correlated
   D. interrelated

9. The _____ is a factor the researcher believes causes changes in the dependent variable.
   A. correlated variable
   B. interrelated variable
   C. independent variable
   D. hypothesis

10. Which of the following research methods describes an experimental situation in which the researcher observes and studies subjects in their natural setting?
    A. Participant observation
    B. Field experiment
    C. Survey
    D. Controlled experiment

11. What process guides the sociologist doing quantitative research?
    A. Interview
    B. Survey
    C. Scientific method
    D. Social method

12. According to Firestone and Harris' research findings, _____ is one of the means of keeping stratification strong within the military.
    A. participatory organization
    B. special training
    C. sexual harassment
    D. hierarchy of power

13. After reviewing the literature, _____ is the next step in the scientific method.
    A. formulating research questions
    B. forming conclusions
    C. reporting findings
    D. sharing information

14. Sociologists Juanita Firestone and Richard Harris designed their research study to use _____ analysis.
    A. quantitative
    B. qualitative
    C. subjective
    D. objective

15. When Martin Sanchez-Jankowski spends months with youth gangs, gaining their trust and learning their norms, he is using the _____ research method.
    A. participant observation
    B. field experiment
    C. survey
    D. controlled experiment

16. Letizia Battaglia and Franco Zeechin used _____ to create a photographic record of violence and Mafia influence in the Sicilian town of Palermo in 1989.
    A. interviews
    B. informal observation
    C. visual sociology
    D. review of old photographs

17. The term _____ effect is used to refer to any unintended effect that results from attention given to subjects in an experiment.
    A. *halo*
    B. *Hawthorne*
    C. *bias*
    D. *attention*

18. Sociologist John Gardner wanted to find out _____.
    A. how many people participated in blue grass music
    B. how and why people are drawn to blue grass music
    C. what people think about blue grass music
    D. how to participate in blue grass music

19. Sociologists William Kornblum and Terry Williams used a _____ method for studying _____ in New York City.
    A. qualitative; homelessness
    B. quantitative; homelessness
    C. mixed; homelessness
    D. mixed; poverty

20. Which of the following is a hypothesis that guided Kornblum and Williams' research?
   A. When you begin to meet the needs of the homeless population, you can begin to move them into socially constructive roles for themselves and society.
   B. As the number of homeless inside the homeless center increases, the number of homeless on the streets decreases.
   C. Both of the above
   D. None of the above

Essay Question
Answer the following question using complete sentences in a well-developed essay.

1. What makes the information gained through sociological research different from information you read in everyday newspapers and magazines? Provide specific examples from newspapers and magazines that illustrate these differences.

ANSWER KEY

The following provides the answers and references for the practice test questions.

Multiple Choice:
1. D..............LO 1..textbook, pp. 24–25
2. D..............LO 1..video segment 2
3. A..............LO 1..textbook, p. 24; video segment 2
4. C..............LO 1..video segment 2
5. B..............LO 1..video segment 2
6. D..............LO 1..video segment 2
7. B..............LO 2..textbook, pp. 25–26
8. A..............LO 2..textbook, p. 25
9. C..............LO 2..textbook, p. 25
10. B..............LO 2..textbook, pp. 29, 31
11. C..............LO 2..video segment 3
12. C..............LO 2..video segment 3
13. A..............LO 2..textbook, p. 24; video segment 3
14. A..............LO 2..video segment 3
15. A..............LO 3..textbook, p. 29
16. C..............LO 3..textbook, p. 29
17. B..............LO 3..textbook, pp. 31–32
18. B..............LO 3..video segment 4
19. C..............LO 4..video segment 5
20. C..............LO 4..video segment 5

Essay Question

Answers should include the following types of statements/points:

1. ....LO 1 ....textbook, pp. 24–39; video segments 1–6
   <u>Sociological research</u>
   - Objectivity
   - Research-based
   - Reliability
   - Validity
   - Identifies source(s) of data
   - Often employs statistical methods of measurement
   - Helps explain relationship of variables
   - Interpretation of data helps develop "plans" of action "to alleviate problems"

   <u>Magazines and Newspapers</u>
   - May contain bias
   - Does not give data to support information
   - Often contain "sensationalized" information
   - Much material based on what people "want" to read
   - Objective is to sell magazines and newspapers
   - Presents incomplete information
   - Often does not credit source of information or uses "confidential source"

LESSON CONTRIBUTORS

Victor A. Ayala, Professor of Sociology, New York City Technical College, Brooklyn, NY

Juan Battle, Professor of Sociology, Hunter College, New York, NY

Kay Coder, Professor of Sociology, Richland College, Dallas, TX

Juanita Firestone, Professor of Sociology, University of Texas at San Antonio, San Antonio, TX

Robert O. Gardner, Sociologist/Instructor, University of Colorado—Boulder, Boulder, CO

William Kornblum, Professor of Sociology, City University of New York, New York, NY

Terry Williams, Associate Professor, New School University, New York, NY

Special acknowledgement and thanks go to:
   Grand Central Neighborhood Social Services Corporation, New York, NY

Lesson 4

Culture

OVERVIEW

Culture is a basic concept in sociology because it is what makes humans unique. Some people think of culture as the fine arts (music, operas, ballets, theatre). To the sociologist, culture is used in a more encompassing way to include the entire way of life of a people. All members of society are *cultured*.

The American culture represents dominant ways Americans exhibit distinctive language, food, dress, beliefs, and behavior patterns. Yet, within this large country, there are distinctive and unique ways people in groups exhibit their own language, food, dress, beliefs, and behavior patterns.

The rodeo is a subculture that has its roots in the Old West. Many of the traditions and customs of rodeo are shared by the members who participate in rodeo life.

In this lesson, a college student talks about his experiences after coming to the United States. It took him several months to begin to learn the cultural traits that make America so unique. But it did not come naturally. He often wondered *why* things happen as they do; *why* the eating patterns and diet of Americans are the way they are; *why* dating practices are the way they are. He often felt like a "fish out of water"—but he quickly learned how to fit in yet retain his own individuality and unique cultural customs and traditions.

It has been said that our world is shrinking – becoming smaller as our capabilities to communicate and travel worldwide have broadened in such a relatively short time. As we expand our professional and personal worlds, why is it so important to become culturally sensitive and aware of others' ways of life? Why must we not be so quick to judge the way others live? How do we develop an ability to gain cultural understanding of people of throughout the world?

LESSON ASSIGNMENTS

Text: William Kornblum: *Sociology in a Changing World*, Chapter 3, "Culture," pp. 44–70

Video: "Culture" from the series *Exploring Society, Introduction to Sociology*

LESSON GOAL

After completing this lesson, you will comprehend the dimensions of culture and appreciate the significance of culture to society.

LESSON LEARNING OBJECTIVES

1. Using an example, discuss each of the dimensions of culture.
2. Explain the significance of the sociobiological hypothesis.
3. Discuss the aspects of cultural understanding.
4. Explain how civilizations are related to cultural changes.
5. Using an example, discuss the aspects of subculture.

REVIEW

The following steps are intended to help you learn the material in this lesson. To maximize your learning experience:

a. Scan the Lesson Focus Point questions.
b. Read the assigned text pages.
c. View the video.
d. Write answers to the Lesson Focus Point questions.
e. Complete the Related Activities assigned by your instructor. If none are assigned, use them to help you review the lesson material.
f. Take the Practice Test and check your answers with the Answer Key located at the end of the lesson.

LESSON FOCUS POINTS

1. From a sociological perspective, why is culture considered multi-dimensional?
2. How does sociology define the three dimensions of culture?
3. What are some present-day examples of each dimension of culture?
4. Which dimensions of culture have multiple levels, and why?
5. What are some examples of the multiple levels in a dimension of culture?
6. What is *social control*? How are norms related to social control?
7. What are *sanctions*?
8. What are the different levels of norms?

9. What is a *formal norm*? What is an *informal norm*?
10. What is *cultural evolution*? *Social Darwinism*?
11. What is *sociobiology*?
12. Who is Jane Goodall?
13. What are *attitudes*?
14. How do attitudes predisposition us to think, feel, or act in a certain way?
15. Why do attitudes have a cultural component?
16. Why do we tend to be very protective of our attitudes?
17. If attitudes become ethnocentric in nature (i.e., even patriotism can be considered ethnocentric), how do they affect our ability to experience and/or participate in other cultures?
18. Why is language so important to culture?
19. What is the linguistic-relativity hypothesis?
20. How does a shared understanding of language and the ability to suspend judgement about other cultures (cultural relativity) enable us to participate in each other's culture?
21. What are the positive aspects associated with crossing cultural lines?
22. What is *cultural hegemony*?
23. What is a *civilization*? How is culture related to civilization?
24. What are the effects of cultural contact? Give examples.
25. What is a *subculture*?
26. What is a *counterculture*?

RELATED ACTIVITIES

1. Although it is not accurate to say each of us has our own culture, each of us lives in a society, community, or family that surrounds us with culture. In general terms, characterize your culture, that is, the culture with which you identify the most. In short, describe your "cultural heritage."

2. Describe three norms you accept and believe in. In your sentences, use normative words, such as should, must, ought. Example: One should say thank you when given a gift. Identify whether they are folkways, mores, or laws.

3. In single words, make a list of four of your values. Below your list, identify each of your words for these values by describing how each does and/or does not match traditional American values.

4. Give five examples of laws, customs, or norms that are part of your culture. What is the value represented by each? Describe how each provides a sense of stability or comfort in the face of changing conditions. For example: People on a submarine, on a ship, in a remote location, or in the military can constitute a culture of their own. What laws, customs, or norms are parts of that culture?

5. List and describe words, dialect, or grammatical constructions that you use or have used that reflect your cultural heritage. Perhaps your family is from a social class, or ethnic, religious, or other subculture, that is different from the dominant culture. Even if you're not, each family has its own unique words or ways of saying things that may not match conventional language. Identify these language differences.

6. Describe a situation in which you had difficulty communicating with someone because of language problems. Perhaps the other person spoke a language foreign to you or used a vocabulary with which you were unfamiliar.

PRACTICE TEST

Multiple Choice
Choose the BEST answer.

1. Which of the following is NOT a dimension of culture?
   A. Ideas
   B. Norms
   C. Material culture
   D. Judgements

2. _____ include our accepted ways of doing things.
   A. Norms
   B. Ideas
   C. Practices
   D. None of the above

3. _____ comprise two dimensions of culture—ideas and norms.
   A. Values
   B. Technology
   C. Material culture
   D. Ideologies

4. Probably the most significant of inventions made possible by culture is _____ .
   A. tools
   B. machines
   C. language
   D. computer technology

5. One of the most popular technologies today is _____ .
   A. microwavable foods
   B. bottled water
   C. organic food
   D. cell phones

6. Theories of crime proposing that there are genes that produce criminal behavior is an example of _____ .
   A. biological determinism
   B. genetic predisposition
   C. biological reductionism
   D. biocriminology

7. The _____ hypothesis states that all human behavior is determined by _____ factors.
   A. sociobiology; social
   B. sociobiology; genetic
   C. social-genetic; genetic
   D. social-genetic; genetic and social

8. The European explorers' assumption that the American Indians could benefit from the adoption of European cultural traits, is an example of _____ .
   A. cultural relativity
   B. ethnocentrism
   C. self-serving culturalism
   D. none of the above

9. Recognizing that all cultures develop unique ways of dealing with the demands of their environments is an aspect of _____ .
   A. cultural relativity
   B. ethnocentrism
   C. culturalism
   D. judgement suspension

10. Using the term *American* to refer to citizens of the United States and not those of Canada and South American nations is an expression of _____.
    A. cultural relativity
    B. judgement
    C. ethnocentrism
    D. centralism

11. Indian college student, Sundar Victor, experienced culture shock when he came to America and discovered _____.
    A. the fast pace of life
    B. the difficulty in getting around if you don't have a car
    C. the existence of some of America's nightlife
    D. all of the above

12. One experience that Victor Sundar had related to language was his inability to understand _____.
    A. slang
    B. English
    C. music
    D. religion

13. One of the ways Troy Poole experiences Sundar Victor's culture is through sharing _____.
    A. clothes
    B. video tapes
    C. food
    D. athletics

14. When people from one civilization incorporate norms and values from other cultures into their own, _____ has taken place.
    A. reculturation
    B. inculturation
    C. blended culturation
    D. acculturation

15. Which of the following is NOT a form of cultural contact discussed in your textbook?
    A. Centralization
    B. Acculturation
    C. Accommodation
    D. Assimilation

16. Subcultures may retain their own _____.
    A. language
    B. rituals
    C. tools
    D. all of the above

17. Rodeo as a subculture holds true to traditions founded in the _____.
    A. American West
    B. rural south
    C. eastern seaboard
    D. northern plains

18. The _____ is a general set of values that is shared among rodeo participants.
    A. cowboy code
    B. oath of honesty
    C. rodeo preamble
    D. all of the above

19. An example of how a subculture can have its own language is the word "pawing," which, in the rodeo subculture means _____.
    A. a horse counts with its front leg
    B. the bareback rider straps in
    C. a bull rider is thrown
    D. a bull jumps off the ground with all four feet

20. Many of the tools and equipment used in rodeo come from _____.
    A. cattle ranching
    B. scientific research
    C. individual participants
    D. none of the above

Essay Questions
Answer the following questions using complete sentences in a well-developed essay.

1. Describe a situation in which you had difficulty communicating with someone because of language problems. Perhaps the other person spoke a language foreign to you or used a vocabulary with which you are unfamiliar.

2. List five norms or values you strongly believe in that have helped you to become a participating member of society. Choose from those norms or values that parents, teachers, or "significant others" taught you as you grew up. Select one of the norms and explain how your belief in it has contributed to your becoming an accepted member of your community.

ANSWER KEY

The following provides the answers and references for the practice test questions.

Multiple Choice:

1. D..............LO 1 ....textbook, p. 46
2. A..............LO 1 ....textbook, pp. 46–47
3. D..............LO 1 ....textbook, pp. 46–47
4. C..............LO 1 ....textbook, p. 58
5. D..............LO 1 .... video segment 2
6. C..............LO 2 ....textbook, p. 56
7. B..............LO 2 ....textbook, pp. 56–57
8. B..............LO 3 ....textbook, pp. 59–60
9. A..............LO 3 ....textbook, p. 60
10. C..............LO 3 ....textbook, pp. 59–60
11. D..............LO 3 .... video segment 3
12. A..............LO 3 .... video segment 3
13. C..............LO 3 .... video segment 3
14. D..............LO 4 ....textbook, pp. 64–67
15. A..............LO 4 ....textbook, pp. 63–68
16. D..............LO 5 .... video segment 4
17. A..............LO 5 .... video segment 4
18. A..............LO 5 .... video segment 4
19. D..............LO 5 .... video segment 4
20. A..............LO 5 .... video segment 4

Essay Questions

Answers should include the following types of statements/points:

1. ....LO 2 ....textbook, p. 58–59
   Based on individual experiences. It is interesting to note that when we have communication problems, we often turn to:
   - Speaking more loudly or slowly.
   - Using gestures.
   - Pointing to objects.

   Cultural differences also promote misunderstanding. Often, white students report difficulty understanding African-Americans' "street-language" or "jive."

Rather than asking the person to explain, the student becomes embarrassed and will leave—not understanding.

2. ....LO 1 .... textbook, pp. 46–50 and 51–52

- Voting
- Religion
- Family

- Education
- Physical fitness
- Manners

- Cleanliness

Explanations will vary.

LESSON CONTRIBUTORS

Carol Deming Chenault, Instructor of Sociology, Calhoun Community College, Decatur, Alabama

Kay Coder, Professor of Sociology, Richland College, Dallas, TX

William Kornblum, Professor of Sociology, City University of New York, New York, NY

Michèle Lamont, Professor of Sociology, Princeton University, Princeton, NJ

Mary Patillo, Associate Professor of Sociology, Northwestern University, Evanston, IL

Judith Perrolle, Associate Professor of Sociology, Northeastern University, Boston, MA

Fred Preston, Professor of Sociology, University of Nevada—Las Vegas, Las Vegas, NV

Special acknowledgement and thanks go to:
Adair's Saloon, Dallas, Texas
Cherokee Heritage Center, Tahlequah, OK
Mesquite Championship Rodeo, Mesquite, TX

Lesson 5

Socialization

OVERVIEW

How does a child become a contributing member of society? How does a child learn the cultural values, norms, beliefs, expectations, and rules of society? Can adults learn to become contributing members of society? Do adults learn the cultural values, norms, beliefs, expectations, and rules of society the same way that children do?

In this lesson, you will hear from a psychologist and a sociologist who shed some light on the question of *nature vs. nurture*. Today, most social scientists agree that both nature and nurture contribute to our socialization. It is the socialization process that teaches us how to live in a particular culture. Through the process of socialization, society is able to maintain itself by transmitting culture from one generation to the next.

You will see how the socialization process works within the Dare family. Socialization is always taking place through our experiences in schools, television, church, family outings, and even through our recreational adventures at the video arcade.

Finally, it is critical to view the socialization process as a lifelong process – one that continues from birth to death. There are many sociologists and psychologists who think socialization actually begins *before* birth. While most of the socialization research has focused on the early years of development, today sociologists are now examining the later years of adulthood.

LESSON ASSIGNMENTS

Text: William Kornblum: *Sociology in a Changing World*, Chapter 5, "Socialization," pp. 97–124

Video: "Socialization" from the series *Exploring Society, Introduction to Sociology*

LESSON GOAL

After completing this lesson, you will comprehend the significance of the socialization process in social development.

LESSON LEARNING OBJECTIVES

1. Discuss the significance of socialization.
2. Identify and discuss the controversial issues that are part of the study of socialization.
3. Explain the role of the social environment in the social construction of self.
4. Discuss the roles of agencies of socialization in the social construction of self.
5. Explain why socialization is a life-long process.

REVIEW

The following steps are intended to help you learn the material in this lesson. To maximize your learning experience:

a. Scan the Lesson Focus Point questions.
b. Read the assigned text pages.
c. View the video.
d. Write answers to the Lesson Focus Point questions.
e. Complete the Related Activities assigned by your instructor. If none are assigned, use them to help you review the lesson material.
f. Take the Practice Test and check your answers with the Answer Key located at the end of the lesson.

LESSON FOCUS POINTS

1. What is *socialization*?
2. How do sociologists and psychologists view socialization differently?
3. What are the primary (family) and secondary (peers) effects on social development and the formulation of our social reality?
4. What is the key point of the "nature vs. nurture" debate?
5. What did Sigmund Freud contribute to our understanding of socialization?

6. What is *behaviorism*?
7. What is the significance of the case of Genie?
8. What did Charles Horton Cooley contribute to our understanding of the social construction of *self*?
9. What is the *looking glass self*?
10. What did Mead say about the significance of role-taking?
11. What is *face*?
12. What contributions have been made by Jean Piaget, Lawrence Kohlberg, and Carol Gilligan?
13. What is an *agency of socialization*?
14. What are some current examples of each of the agencies of socialization?
15. What is the role of each agency in the socialization process?
16. How do the agencies of socialization vary in their weight of influence on an individual's socialization process?
17. How does/could change within agencies of socialization affect an individual's socialization process, with emphasis given to the mass media (i.e., increasing TV violence) and the family (i.e., increasing numbers of working parents and need for employer provided day care)?
18. How/why does the socialization process and the social construction of "self" continue throughout a person's lifetime?
19. What does adult socialization mean?
20. Why are sociologists concerned with adult socialization?
21. Why, in adulthood, is our socialization usually more apparent when a crises of some type affects our lives—like divorce, birth of a child, death of parents or loved ones?
22. What is resocialization?
23. What are some present-day examples of resocialization?
24. Where does resocialization usually take place?
25. What are total institutions?
26. What stages are parts of Erickson's lifelong socialization?
27. What are the peer groups identified by a journalist studying peer groups n high schools?

RELATED ACTIVITIES

1. Describe a situation in which your parents or guardians interacted with you and taught you a social skill that you have found to be helpful in your life. Why was it important to your life?

2. List and explain three social skills you learned from your parent(s); how did you learn each skill?

3. On the job—at work—you acquire knowledge, skills, and values, all of which help you to be a participating member of society. Briefly describe at least five things you have learned from those with whom you have worked that have increased your ability to participate in society.

4. The "looking glass self" has to do with the degree of self-esteem you have, based on your perceptions of the judgments of others about you. Some of these perceptions correspond to the actual evaluations of others; some do not. Make a list of two or three of your significant others—either dead or alive—and, to the right of their names, describe what *you think* their assessments of you are. For each case, describe the reasons for your assessment.

5. Describe five things you learned from your schools that have helped you to survive in your society and to participate effectively in your community. List at least two things that could have helped you that you regret you did not learn at school. Explain why you feel this way.

6. Peer groups have both positive and negative influences on us. Explain how your peers have both helped and harmed you as a person.

7. Think of books, magazines, or other reading materials you were exposed to as a child. Analyze the content of those reading materials in terms of gender socialization. What pictures, photos, words, or examples perpetuated in you the idea of separate gender roles? Summarize your findings. If you can, describe ways that you took on the gender roles presented in the materials.

PRACTICE TEST

Multiple Choice
Choose the BEST answer.

1. _____ is the process by which we learn to conform to society's norms.
   A. Culture
   B. Socialization
   C. Conformity
   D. Environment

2. Socialization is a _____ process.
   A. childhood
   B. dysfunctional
   C. life long
   D. early life

3. Which of the following would NOT contribute to our socialization?
   A. Family
   B. School
   C. Clergy
   D. None of the above

4. The nature versus nurture debate has primarily taken place between sociologists and _____.
   A. environmentalists
   B. psychologists
   C. historians
   D. anthropologists

5. The *nurture* side of the debate deals primarily with _____factors.
   A. social
   B. psychological
   C. physical
   D. genetic

6. The early sociologists _____ and _____ emphasized the *nurture* side of the *nature vs. nurture* debate.
   A. Charles Cooley; Herbert Mead
   B. Sigmund Freud; Darwin
   C. Charles Cooley; Sigmund Freud
   D. All of the above

7. The way people treat you greatly influences your _____ concepts.
   A. biological
   B. genetic
   C. social
   D. all of the above

8. Studies of children reared in extreme isolation have pointed researchers to the suggestion that lack of parental attention can result in _____.
   A. retardation
   B. death
   C. emotional problems
   D. all of the above

9. Which of the following reasons do social scientists use to counter the conclusions reached by Richard Herrnstein and Charles Murray?
   A. Intelligence is far too complex to be represented by a single measure such as IQ.
   B. There is evidence of cultural and middle class biases used in questions to test IQ.
   C. Correlation is not the same as causality.
   D. All of the above.

10. The self is viewed as a social construct. This means it is _____.
   A. constructed through interactions with other people over a lifetime
   B. not based at all on genetic traits
   C. doesn't matter what you are born with but what happens after
   D. all of the above

11. The _____ is the reflection of our self that we think we see in the behaviors of others around us.
    A. significant other
    B. role assumption
    C. looking glass self
    D. mirrored image

12. The people whose behavior we imitate are called _____ others.
    A. significant
    B. primary
    C. secondary
    D. generalized

13. Role _____ refers to the way we try to look at social situations from the standpoint of another person from whom we seek a response.
    A. assumption
    B. taking
    C. playing
    D. acting out

14. Which of the following is NOT a stage in the development of role taking?
    A. Play
    B. Anticipatory
    C. Preparatory
    D. Game

15. _____ is a process whereby individuals undergo intense, deliberate socialization designed to change major beliefs and behaviors.
    A. Secondary socialization
    B. Rehabilitation
    C. Generalized socialization
    D. Resocialization

16. Which of the following is the primary agency of socialization?
    A. Schools
    B. Peer group
    C. Mass media
    D. Family

17. Which of the following would NOT be a community agent of socialization?
    A. Day care centers
    B. Scout troop
    C. Recreation centers
    D. Schools

18. Which of the following agencies of socialization is the most dominant in middle and late childhood?
    A. Family
    B. School
    C. Peers
    D. Community groups

19. In studying different peer groups among adolescents, which statement is NOT true?
    A. Adolescents acquire much of their identity from their peers.
    B. Adolescents find no difficulty in deviating from their peer groups.
    C. Peer groups may becomes more important than family.
    D. There is often conflict in a family caused by the influence of peer groups.

20. Two of the influences that produce socialization after childhood are _____ and _____.
    A. significant others; occupational mobility
    B. school; church affiliation
    C. mass media; occupation
    D. significant others; mass media

Essay Questions
Answer the following questions using complete sentences in a well-developed essay.

1. List and explain three social skills you learned form your parent(s); how did you learn each skill?

2. What kind of rewards and positive reinforcements were you given as a child by your significant others to help shape your behavior? What punishments were used? Which of them were appropriate, and which were abusive, if any?

ANSWER KEY

The following provides the answers and references for the practice test questions.

Multiple Choice:

1.	B	LO 1	textbook, p. 98; video segment 2
2.	C	LO 1	textbook, p. 98; video segment 2
3.	D	LO 1	video segment 2
4.	B	LO 2	video segment 2
5.	A	LO 2	textbook, pp. 98–99; video segment 2
6.	A	LO 2	textbook, pp. 105–107; video segment 2
7.	C	LO 2	video segment 2
8.	D	LO 2	textbook, pp. 101–102
9.	D	LO 2	textbook, pp. 102–103
10.	D	LO 3	textbook, p. 105
11.	C	LO 3	textbook, p. 105
12.	A	LO 3	textbook, p. 106
13.	B	LO 3	textbook, p. 105
14.	B	LO 3	textbook, pp. 105–107
15.	D	LO 3	textbook, pp. 117–118
16.	D	LO 4	textbook, p. 112
17.	D	LO 4	textbook, pp. 114–115
18.	C	LO 4	textbook, p. 116
19.	B	LO 4	textbook, p. 116
20.	A	LO 5	textbook, p. 117

Essay Questions

Answers should include the following types of statements/points:

1. ....LO 4...textbook, pp. 112–114

- My parents taught me to give up my seat on a bus (or any transportation) to an older person out of respect. Because I was younger, it was easier for me to stand.
- They taught me this through verbal reinforcement. They also taught me to open doors for anyone behind me—males and females. This is an act of courtesy, and when I did not do it when younger, they corrected my actions through verbally reinforcing my actions when performed correctly.
- I learned "speaking" skills through both my parents' emphasis on "good" language usage (oral and written) and my teachers' reinforcements when I demonstrated good language usage (oral and written.) The biggest reinforcement I received came in the form of good grades on oral and written reports.

2. ....LO 3, 4 ....textbook, pp. 112–114

- Rewards/Positive Reinforcements: money, touch, hug, gift, verbal praise
- Punishments: withdrawal of reward, grounding, yelling, spanking, ignoring, use of profanity
- Appropriate: (based on individual's experience)
- Abusive: (based on individual's experience)

LESSON CONTRIBUTORS

Wanda Alderman-Swain, Professor of Sociology, Howard University, Washington, DC

Bruce A. Chadwick, Professor of Sociology, Brigham Young University, Provo, UT

Arlie Russell Hochschild, Professor of Sociology, University of California—Berkeley, Berkeley, CA

Theresa A. Martinez, Associate Professor of Sociology, University of Utah, Salt Lake City, UT

Farnoosh Massoudian, Psychology Instructor, Brookhaven College, Farmers Branch, TX

Special acknowledgment and thanks to:
  Darla Adams, Garland, TX
  Joan Adams, Garland, TX
  The Dare Family, Garland, TX
  Olivia Figueroa, Garland, TX
  Travis Figueroa, Garland, TX
  Randall Ford, Garland, TX
  Beaver Technology Center, Garland, TX
  Garland High School, Garland, TX

Lesson 6

Social Interactions, Relationships, and Structures

OVERVIEW

Sociologists seek to understand social interaction as it occurs in everyday activities. A telephone conversation, playing football on a Saturday afternoon, or having lunch with business associates—these are all activities in which our actions are greatly influenced by the actions of the people around us.

The social interaction patterns we establish among people, which include the spoken and unspoken expectations of everyday interaction, pave the way for the establishment of more lasting social connections. These social connections are called relationships. Relationships take many different forms and establish the foundation of our social world.

In this lesson, the participants of a ropes course learn the importance of working together to accomplish tasks. Throughout the United States, work groups participate in such training programs to learn the importance of team effort in achieving a common goal.

The sociological concept of social structure implies a pattern of regularity to the social settings in which individuals interact to form relationships. Social structure includes the social institutions, organizations, groups, statuses and roles, values, and norms that give us order and predictability. It would be impossible to formulate meaningful interpretations of human behavior if social structure happened by chance alone.

The U.S. Naval Academy is strong in tradition and rich in history. It offers us an illustration of well-defined relationships between individuals and groups. There is little room for confusion—the social structure is clearly stipulated from the moment students of the Academy become members of the Naval Brigade.

But the military isn't the only illustration of the importance and impact of social structure. The family offers an example of a social structure in which family members interact in specified ways on a regular basis. Parents are expected to assume financial obligations for children, and children are expected to obey their parents. Within a family, the social structure is established by the family members. For example, most family members are not formally assigned a specific place to sit at the dinner table. But let one of the children sit in someone else's place, and the effects of social structure are immediately apparent!

LESSON ASSIGNMENTS

Text: William Kornblum: *Sociology in a Changing World*, Chapter 4, "Societies and Nations," pp. 72–77, and Chapter 6, "Interaction in Groups," pp. 134–137 and pp. 140–143

Video: "Social Interactions, Relationships and Structures" from the series *Exploring Society, Introduction to Sociology*

LESSON GOAL

After completing this lesson, you will comprehend the role of social interaction and the effects of social structure in establishing and maintaining relationships.

LESSON LEARNING OBJECTIVES

1. Explain the four principles of interaction.
2. Discuss the elements of social structure.
3. Using an example, explain how group leaders emerge from group interaction.

REVIEW

The following steps are intended to help you learn the material in this lesson. To maximize your learning experience:

a. Scan the Lesson Focus Point questions.
b. Read the assigned text pages.
c. View the video.
d. Write answers to the Lesson Focus Point questions.
e. Complete the Related Activities assigned by your instructor. If none are assigned, use them to help you review the lesson material.
f. Take the Practice Test and check your answers with the Answer Key located at the end of the lesson.

LESSON FOCUS POINTS

1. What is the *pleasure principle* of interaction? Give examples.
2. What is the *rationality principle* of interaction? Give examples.
3. What is the *reciprocity principle* of interaction? Give examples.
4. What is the *fairness principle* of interaction? Give examples.
5. How can the basic principles of interaction—pleasure seeking, rationality, reciprocity, and fairness—be used to explain the emergence of group structure?
6. What contributions did the research by George Homans make to our understanding of group structure?
7. How does social interaction contribute to group leadership?
8. What is the significance of a task leader and an emotional leader?
9. How does the ropes course illustrate emergence of group leaders?
10. What is *social structure*? (See definition on page 93.)
11. What is a *group*? Give examples.
12. What is *status*? Give examples.
13. What is a *role*? Give examples.
14. What are *role expectations*? Give examples.
15. How do the *elements of social structure* form the recurring patterns of behavior that people create through their interactions, their exchanges of information, and their relationships?
16. What is the relationship of social interaction to social structure?
17. How does a kinship chart provide a visual model of the social structure of a family?
18. How does social change affect the social structure of groups?
19. What was revealed by the research of William F. Whyte (1984) regarding the roles of successful waitresses?
20. What happens to social structure when you put people together?
21. What happens to social structure any time you change the players (such as on the popular television series, "*Survivor*," when someone is "voted" off or wins "immunity")?

RELATED ACTIVITIES

1. Describe a role conflict you experience in your life, including the feelings that accompany the stresses involved. In your narrative, identify the two (or more) roles that are in conflict.

2. Role strain is something most of us go through in modern society. Explain how the strain of trying to meet contradictory demands or new expectations in a new role affects you. Identify the role and two or more demands or expectations involved.

3. Often we are not comfortable or happy with our ascribed statuses. Describe ways in which you have had difficulty, discomfort, or unhappiness with one of your ascribed statuses.

4. Using yourself or another person as the example, describe ways in which a master status occasioned prejudice or discrimination. (Master status can be occupational status as well as gender, race, ethnicity, age, or physical appearance.)

PRACTICE TEST

Multiple Choice
Choose the BEST answer.

1. Which of the following is NOT one of the four principles of interaction?
   A. Pleasure principle
   B. Rationality principle
   C. Reciprocity principle
   D. Equity principle

2. When we socialize with people who provide support and acceptance, the _____ principle is at work.
   A. rationality
   B. selective
   C. pleasure
   D. reciprocity

---

3. The _____ principle states that people change their behavior according to whether they think they'll be worse or better off as a consequence.
   A. fairness
   B. rationality
   C. reciprocity
   D. pleasure

4. Asking the question, "What do I receive out of this interaction?" is an example of the _____ principle.
   A. fairness
   B. rationality
   C. pleasure
   D. reciprocity

5. In the Mayo studies in the Hawthorne plant, the workers arrived at their own definition of what they considered _____.
   A. cost efficient
   B. fair worker output
   C. equitable
   D. all of the above

6. The four principles of interaction work together to form _____.
   A. group principles
   B. group structure
   C. group parameters
   D. all of the above

7. Which of the following is an example of a status?
   A. Practical nurse
   B. Registered nurse
   C. Chief resident
   D. All of the above

8. Which of the following is NOT an element of social structure?
   A. Group
   B. Role
   C. Organization
   D. Status

9. Social _____ refers to the recurring patterns of behavior that people create through their social interactions.
    A. group
    B. structure
    C. status
    D. role

10. A kinship diagram is one way sociologists and anthropologists illustrate social _____.
    A. structure
    B. groups
    C. statuses
    D. roles

11. The way a society defines how an individual is to behave in a particular status is referred to as a _____.
    A. role
    B. status expectation
    C. role expectation
    D. social expectation

12. Which of the following is NOT an example of a status?
    A. Father
    B. Mother
    C. Son
    D. School

13. In sociology, the term institution is used to designate _____.
    A. a large bureaucratic organization
    B. stable structures of statuses and roles devoted to meeting the basic needs of people in society
    C. an organization such as a prison
    D. none of the above

14. Social scientists have identified certain principles of interaction that helps explain _____ and _____ in human groups.
   A. group behavior; group interaction
   B. stability; change
   C. individual behavior; motivation
   D. motivation; interactions

15. A landlord rents out a small house. When the lease is due for renewal, the landlord learns that the tenant has taken a job very close to the house and is therefore unlikely to move. The landlord raises the rent $40 more than he was planning to. Most people would consider this scenario a violation of the _____ principle.
   A. reciprocity
   B. rationality
   C. pleasure
   D. fairness

16. "There is no duty more indispensable than that of returning a kindness. All men distrust one forgetful of a benefit." This exemplifies the _____ principle.
   A. reciprocity
   B. fairness
   C. rationality
   D. pleasure

17. Within the social structure of the U.S. Naval Academy, Adriel Morgan refers to his group of classmates as _____.
   A. plebites
   B. the mod squad
   C. middies
   D. comrades

18. According to Robert F. Bales and Philip Slater, the leader who emerges in a social group _____.
   A. initiates the most interactions to complete a task
   B. initiates the most interactions to support other members
   C. both A and B
   D. none of the above

19. According to Robert F. Bales and Philip Slater, the person who initiates the second highest number of interactions in a group is the _____.
   A. best-liked person in the group
   B. task leader
   C. primary leader
   D. hardest working member of the group

20. The primary leader is sometimes called the _____ leader.
   A. focus
   B. secondary
   C. emotional
   D. task

Essay Question
Answer the following question using complete sentences in a well-developed essay.

1. If you are employed by a large company, diagram the social structure of the organization using job statuses. Explain what the diagram reveals about the social structure of the organization.

ANSWER KEY

The following provides the answers and references for the practice test questions.

Multiple Choice:
1. D..............LO 1 ....textbook, p. 135–137; video segment 2
2. C ....LO 1 ....textbook, p. 135; video segment 2
3. B ....LO 1 ....textbook, pp. 135–136; video segment 2
4. B ....LO 1 ....textbook, pp. 135–136; video segment 2
5. B ....LO 1 ....textbook, p. 137; video segment 2
6. B ....LO 1 .... video segment 2
7. D..............LO 2 ....textbook, pp. 75–76
8. C..............LO 2 ....textbook, p. 74; video segment 3
9. B..............LO 2 ....textbook, p. 74
10. A..............LO 2 ....textbook, p. 75
11. A..............LO 2 ....textbook, p. 76
12. D..............LO 2 ....textbook, pp. 75–76
13. B..............LO 2 ....textbook, pp. 77 and 79
14. B..............LO 2 ....textbook, p. 135
15. D..............LO 2 ....textbook, pp. 136–137

Essay Question

Answers should include the following types of statements/points:

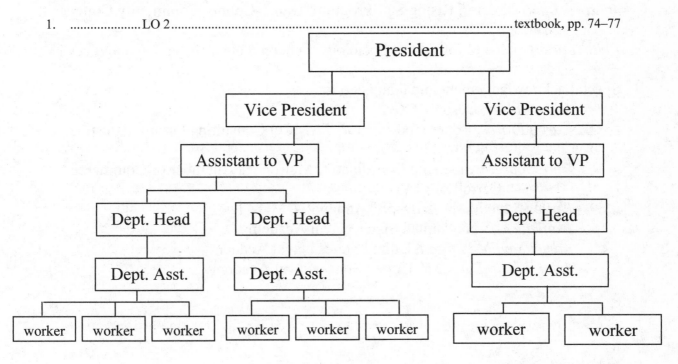

The social structure of this company reflects power and decisions being made by those in the most powerful positions—probably vice presidents and assistant vice-presidents—with department heads having some input. It shows the workers having little power and the social structure does not encourage worker participation.

LESSON CONTRIBUTORS

Tracey McKenzie Elliott, Professor of Sociology, Collin County Community College, Frisco, TX

Juanita Firestone, Professor of Sociology, University of Texas at San Antonio, San Antonio, TX

Leslie Irvine, Assistant Professor of Sociology, University of Colorado—Boulder, Boulder, CO

Julia W. Lam, Director, Rising Star Program, Dallas County Community Colleges, Dallas, TX

Philip Luck, Assistant Professor of Sociology, Lyndon State College, Lyndonvill, VT

Special acknowledgement and thanks go to:

Mary Balser, Desoto, TX

Karen Flores, Ropes Coordinator, TAMCO Consulting Group, Palestine, TX

Rita Moore, Leadership Coordinator, Metrocrest Chamber of Commerce, Carrollton, TX

United State Naval Academy, Annapolis, MD

Danielle LaSalle, United States Naval Academy

Adriel Saville Morgan, United States Naval Academy

John Walsh, Ens USN, United States Naval Academy

Lesson 7

Social Groups

OVERVIEW

We live our lives in social groups. Group membership is central to social life. We are born into, raised, and educated within social groups. Interacting with families, friends, classmates, church groups, and work groups, we learn that we are not the center of the universe but are cells within a larger social organism. Belonging to these groups, we yield to others the right to make certain decisions about our behavior; at the same time, we assume obligations to act according to the expectations of others.

In the video lesson, you will revisit a famous account of bystander apathy that took place in New York City. Consider what you would have done in a similar situation. Then you will see how a similar event occurred since then. What is it about group behavior that makes us think we are not responsible for events which occur around us?

Sociologists observe features that characterize *groups*. One characteristic feature is that most group members share common goals and interests. Your classmates share a common goal of successfully completing the course; a neighborhood watch group shares the common goal and interest of maintaining a safe neighborhood. Through the shared common goal or interest, the group offers its members a sense of belonging to something larger than self or a random collection of people. Such group support and companionship are appealing and influence our thoughts and actions.

In this video lesson, you will meet sociologists who talk about the characteristics of social groups. Then you will see these characteristics illustrated by a high school cheerleading squad. Cheerleading is successful when the group members understand the expectations of all the members; in that way, each person can clearly see the relationship of her/his activity to the activities of the other squad members.

Groups usually have an identifiable structure with rules, expected behaviors, relationships, and perhaps some type of leadership. Such structures can be very loosely defined or more formal. In this video lesson, you will learn about different leadership styles as well as the impact of group behavior—especially as group behavior becomes a type of conformity that ends in blind obedience. Why does this happen?

The interaction we experience in social groups cannot help but shape who we are as individuals, yet at the same time, our unique individual traits and talents influence the groups to which we belong.

LESSON ASSIGNMENTS

Text: William Kornblum: *Sociology in a Changing World*, Chapter 6, "Interaction in Groups," pp. 127–130, pp. 132–134, and pp. 140-145

Video: "Social Groups" from the series *Exploring Society, Introduction to Sociology*

LESSON GOAL

After completing this lesson, you will be able to explain the complexities of social groups and their significance to society.

LESSON LEARNING OBJECTIVES

1. Using an example, discuss the dimensions that define social groups.
2. Discuss the concept of group leadership styles.
3. Examine the consequences of conformity in complex societies.

REVIEW

The following steps are intended to help you learn the material in this lesson. To maximize your learning experience:

   a. Scan the Lesson Focus Point questions.
   b. Read the assigned text pages.
   c. View the video.
   d. Write answers to the Lesson Focus Point questions.
   e. Complete the Related Activities assigned by your instructor. If none are assigned, use them to help you review the lesson material.
   f. Take the Practice Test and check your answers with the Answer Key located at the end of the lesson.

LESSON FOCUS POINTS

1. What is a *social group*? Examples?
2. What are the characteristics of social groups that define them as such? Examples?
3. What is a *primary group* and a *secondary group*? Give examples.
4. What is a *dyad* and a *triad*?
5. What is a *reference group*? Give examples.
6. What effects do different social groups have on the *individual*? Examples?
7. What are the defining characteristics of different leadership styles and followers found in social groups in complex societies? Examples?
8. What consequences for *group* and *leader* do the different leadership styles yield? Examples?
9. Is following orders considered a defense of action?
10. What consequences does the degree of willingness to conform have for *group* and *leader*? Examples?
11. When does conformity become blind obedience?

RELATED ACTIVITIES

1. Name two primary groups to which you belong. Primary groups can sometimes be dysfunctional (harmful) to individuals. Cite one or two instances you know of in which this is the case.

2. Name a secondary group to which you belong or describe a secondary relationship that you have.

3. Explain and give examples of primary and secondary groups.

4. Describe an example of boundary maintenance of a group that you or someone you know belongs to.

5. Name an in-group to which you belong, and describe a situation in which you were excluded from an out-group.

6. Name a reference group to which you belong and have membership, and name a reference group to which you do *not* presently belong.

7. Describe a situation in which you decided *not* to join a group or interact with someone as a friend because the "losses" would be too great.

Multiple Choice
Choose the BEST answer.

1. Which of the following is NOT a dimension of social groups?
   A. Cohesion
   B. Boundaries
   C. Social interaction
   D. Social structure

2. _____ groups are characterized by relationships that involve few aspects of their members personalities.
   A. Primary
   B. Secondary
   C. Intimate
   D. Relationship

3. When a group has _____ members, the relationships among the members become so complex that the group is likely to break into smaller groups.
   A. six
   B. ten
   C. fifteen
   D. all of the above

4. Peers are an example of a(n) _____ group.
   A. in-
   B. out-
   C. reference
   D. primary

5. Which one of the following is NOT a dimension of social groups that can be found in a cheerleading squad?
   A. Membership criteria
   B. Group norms
   C. Specific goals
   D. None of the above

6. Being voted as a captain of the cheerleaders is an example of a _____.
   A. role
   B. status
   C. popularity
   D. ability

7. The cooperation of teamwork needed for a cheerleading squad to be successful is an example of the dimension of _____.
   A. talent
   B. popularity
   C. ability
   D. cohesion

8. _____ are the rules that the cheerleaders follow to maintain cohesion within the social group.
   A. Norms
   B. Statuses
   C. Goals
   D. Roles

9. The story of Kitty Genovese is an example of _____.
   A. bystander effect
   B. crowd dispersion
   C. observers disconnect
   D. bystander anonymity

10. If you were out walking and saw someone you didn't know in need of assistance, research has revealed that you would probably _____.
    A. jump into the situation to render assistance
    B. shy away from involvement if you are in a group
    C. seek help for this person
    D. ignore the situation completely

11. Group leaders are often characterized by their _____ and ability to get tasks completed.
    A. leadership ability
    B. support of suggestions of others
    C. leadership power
    D. leadership authority

12. Which leadership style is most effective when there is a crisis involved?
    A. Democratic
    B. Laissez-faire
    C. Authoritarian
    D. None of the above

13. A _____ leadership style seeks input from group members before making decisions.
    A. laissez-faire
    B. democratic
    C. authoritarian
    D. balanced

14. If a therapist tells a couple, "You know, you need to take responsibility for this session. I'm not going to tell you what to do," which of the following leadership styles is the therapist using?
    A. Democratic
    B. Authoritarian
    C. Laissez-faire
    D. Counseling

15. The experiments of Stanley Milgram were designed to _____.
    A. examine the act of obeying
    B. observe people in their informal groups
    C. test the rule of efficiency in formal organizations
    D. discover the level of emotions under which people operate with greatest productivity

16. Stanley Milgram was dismayed to discover that very large proportions of his subjects were _____.
    A. actively involved in bureaucracies
    B. willing to obey any order given by the experimenter
    C. unwilling to administer the highest level of shock to the experimental subjects
    D. more responsive to informal group pressures than to formal sanctions

17. Which of the following statements about conformity is NOT true?
    A. Conformity is part of our daily lives.
    B. Conformity is necessary in a complex society.
    C. Conformity contributes to stability in society.
    D. Conformity only has positive societal implications.

18. During World War II, Nazi officers who stood trial after the war, said they were simply following orders. This is an example of _____.
    A. groupthink
    B. blind obedience
    C. military might
    D. military anomie

19. Stanley Milgram's research illustrated our willingness to _____.
    A. obey commands of authority figures
    B. conform to our own expectations
    C. counter the arguments of those in authority
    D. let others do our *dirty work*

20. The successful induction into a military unit involves _____.
    A. conformity
    B. obedience
    C. taking away individuality
    D. all of the above

Essay Question
Answer the following question using complete sentences in a well-developed essay.

1. Using examples, explain primary and secondary groups.

ANSWER KEY

The following provides the answers and references for the practice test questions.

Multiple Choice:

1. C ....LO 1 ....textbook, pp. 127–128; video segment 2
2. B ....LO 1 ....textbook, p. 128
3. D ....LO 1 ....textbook, pp. 128–129
4. A ....LO 1 ....textbook, p. 132
5. C ....LO 1 .... video segment 2
6. B ....LO 1 .... video segment 2
7. D ....LO 1 .... video segment 2
8. A ....LO 1 .... video segment 2
9. A ....LO 2 ....textbook, p. 140; video segment 3
10. B ....LO 2 .... textbook, p. 140; video segments 3 and 4
11. B ....LO 2 ....textbook, pp. 141–142; video segment 4
12. C ....LO 2 .... video segment 4
13. B ....LO 2 .... video segment 4
14. C ....LO 2 .... video segment 4
15. A ....LO 3 ....textbook, p. 144; video segment 5
16. B ....LO 3 ....textbook, p. 144; video segment 5
17. D ....LO 3 .... video segment 5
18. B ....LO 3 .... video segment 5
19. A ....LO 3 .... video segment 5
20. D ....LO 3 .... video segment 5

Essay Question

Answers should include the following types of statements/points:

1.   ....LO 1 ....textbook, pp. 128–129

Possible Examples of Primary Groups:
- Family
- Peer groups ("good" friends)
- Support group
- Note: Many students have developed "non-traditional" primary groups (gangs).

Primary Group Characteristics:
- Usually small
- Personal orientation
- Shared activities
- Endures over time
- Time spent together
- Strong loyalty
- Non-interchangeable

Possible Examples of Secondary Groups:
- Organizations
- Associations
- Work group
- Class

Secondary Group Characteristics:
- Relationships involve few aspects of members' personalities
- Joining group based on achieving goals
- Association based on some form of contract on written/unwritten agreement
- Specific scope of interaction
- Size varies

LESSON CONTRIBUTORS

Carol Deming Chenault, Instructor of Sociology, Calhoun Community College, Decatur, Alabama

Gordon Fellman, Professor of Sociology, Brandeis University, Waltham, MA

Phillip B. Gonzalez, Associate Professor, The University of New Mexico, Albuquerque, NM

Theresa A. Martinez, Associate Professor of Sociology, University of Utah, Salt Lake City, UT

Judith A. Perrolle, Associate Professor of Sociology, Northeastern University, Boston, MA

Nathaniel Eugene Terrell, Chairperson of Sociology and Anthropology, Emporia State University, Emporia, KS

Special acknowledgement and thanks go to:
> *Victoria Allen, Chasity Baker, Spencer Koch, Jason Riley*, and the rest of the Garland High School Cheerleading Squad
> *Jennifer Nobra*, Sponsor
> *Garland High School*, Garland, Texas

Lesson 8

Formal Organizations and Bureaucracies

OVERVIEW

Our lives exist in a complex society and are affected by public and private organizations. Such organizations delineate a social structure allowing efficient and predictable ways to accomplish tasks. Such complex systems make bureaucracies necessary. Think about all the things you do during the day related to larger organizational systems. Organizations produce your clothes, your food, your car, the toothpaste you use every morning, the soap you use to wash clothes, the television you watch, the coffee you drink, the newspaper you read, and the computer you use.

Using the sociological perspectives to examine formal organizations and bureaucracies, sociologists help us understand their importance in society and our lives. Efficient production, individuals working together to meet goals, and tensions between management and personnel—these are some of the characteristics revealed by the sociological perspectives.

In this video lesson, you will hear from George Ritzer, author of *The McDonaldization of Society*. Ritzer talks about the effects of bureaucratization on society. According to Ritzer, McDonald's is successful because it represents the classic bureaucracy, with a clear division of labor and a uniform system of rules, making it predictable and highly efficient. For example, you know that at any McDonald's restaurant, the French fries take three minutes and ten seconds to cook and the hamburger patties cook in one minute and forty-eight seconds. An observer once said McDonald's customers "are not in search of 'the best burger they've ever had' but rather 'the same burger they've always had'."

While bureaucracies are efficient structures for managing large groups of people, they are frequently perceived as "iron cages" by those working within them. Some bureaucracies and formal organizations are trying new ways to offset the limitations too often associated with them. In this video lesson, you will learn what one organization, Whole Foods Market, is doing to offset many of the limitations of bureaucracies. You will meet a couple of employees and hear how Whole Foods Market is implementing a new philosophy within this formal organization.

LESSON ASSIGNMENTS

Text: William Kornblum: *Sociology in a Changing World*, Chapter 6, "Interaction in Groups," pp. 143–148

Video: "Formal Organizations and Bureaucracies" from the series *Exploring Society, Introduction to Sociology*

LESSON GOAL

After completing this lesson, you will be able to explain how formal organizations and bureaucracies function.

LESSON LEARNING OBJECTIVES

1. Identify, by example, types of formal organizations.
2. Define and give examples of the characteristics of bureaucracies.
3. Examine bureaucracies and formal organizations using the sociological perspectives.
4. Discuss *The McDonaldization of Society*.
5. Describe the limitations of bureaucracies and formal organizations.

REVIEW

The following steps are intended to help you learn the material in this lesson. To maximize your learning experience:

   a. Scan the Lesson Focus Point questions.
   b. Read the assigned text pages.
   c. View the video.
   d. Write answers to the Lesson Focus Point questions.
   e. Complete the Related Activities assigned by your instructor. If none are assigned, use them to help you review the lesson material.
   f. Take the Practice Test and check your answers with the Answer Key located at the end of the lesson.

LESSON FOCUS POINTS

1. What is the definition of bureaucracy?
2. What are the characteristics of bureaucracies?
3. Why are bureaucracies so interested in conformity?
4. Why is McDonald's so successful? How is that success related to bureaucracy?
5. What is an example of a formal organization?
6. Identify how bureaucracies touch your life.
7. How would a sociologist using the functionalist perspective examine bureaucracies?
8. How would a sociologist using the conflict perspective examine bureaucracies?
9. How would a sociologist using the interactionist perspective examine bureaucracies?
10. What are the limitations of bureaucracies and formal organizations?
11. How is Whole Foods Market trying to overcome the limitations of bureaucracies and formal organizations?

RELATED ACTIVITIES

1. Create an organization which attempts to minimize the negative characteristics of bureaucracies.

2. List and explain three advantages and three disadvantages of working within a large bureaucracy.

3. Describe a situation or incident in which you have experienced two or more typical characteristics of bureaucracy. Explain how these were beneficial and helped you achieve your goals.

4. Describe a situation or incident in which you experienced the dysfunctional characteristics of a bureaucracy.

5. List at least one formal organization that touched your life you during each of these periods: early childhood, elementary school, high school, and college years.

PRACTICE TEST

Multiple Choice
Choose the BEST answer.

1. _____ are groups whose norms and statuses are generally agreed upon but are not written.
   A. Associations
   B. Informal organizations
   C. Formal organizations
   D. None of the above

2. A tenant's association of an urban apartment building is an example of a(n) _____.
   A. formal organization
   B. informal organization
   C. voluntary organization
   D. common association

3. The PTA or Rotary Club is an example of a(n) _____.
   A. informal organization
   B. voluntary association
   C. voluntary organization
   D. interest association

4. _____ is a formal organization whose members pursue shared interests and arrive at a decision through a democratic process.
   A. Bureaucracy
   B. Voluntary association
   C. Involuntary club
   D. Corporation

5. In _____ organizations, members participate in order to attain material rewards and earn a living.
   A. normative
   B. utilitarian
   C. coercive
   D. voluntary

6. A prison would be an example of a _____ organization.
   A. normative
   B. coercive
   C. utilitarian
   D. isolated

7. There are explicit, and often written, rules defining who may participate and the scope and manner of that participation in _____.
   A. formal organizations
   B. informal organizations
   C. primary groups
   D. friendship cliques

8. Which of the following is NOT a characteristic of bureaucratic organizations as described by Max Weber?
   A. Career ladder
   B. Personal relationships
   C. Norm of efficiency
   D. Rules and precedents

9. Max Weber believed that bureaucracy made human social life more _____.
   A. disorganized
   B. rational
   C. irrational
   D. personal

10. According to Max Weber, which of the of the following is NOT a way bureaucracies "rationalize" human societies?
    A. Democracy
    B. Impersonality
    C. Norm of efficiency
    D. Rules

11. A _____ means that promotions are based on seniority or achievement, or both.
    A. bureaucracy
    B. hierarchy
    C. career ladder
    D. regulation

12. A bureaucracy is a specific structure of statuses and roles in which the power to control actions of others is _____.
    A. held by the majority
    B. distributed equally among various levels of the organization
    C. increased as one nears the top of the organization
    D. used to preserve the democratic structure of the organization

13. In their study of the Nazi army, Morris Janowitz and Edward Shils discovered that the source of the soldiers' commitment to the army was the _____.
    A. primary group to which they belonged
    B. bureaucracy that cared for them
    C. ideology they encountered
    D. effective use of propaganda

14. In their study of German soldiers, Morris Janowitz and Edward Shils found that soldiers would continue to fight even in the face of defeat because of their _____.
    A. fear of the wrath of their superiors
    B. commitment to Nazi ideology and their belief in its superiority
    C. excellent training and superior self-discipline
    D. loyalty to their small combat units and their devotion to the other members of the units

15. Which sociological perspective would conclude that parts of the bureaucracy that are not working need to be weeded out?
    A. Interactionism
    B. Conflict
    C. Functionalism
    D. Feminism

16. Which of the following is an example of what sociologists using the functionalist perspective would call an efficiently operated bureaucracy?
    A. Military
    B. Prison system
    C. Police system
    D. All of the above

17. A sociologist using the _____ perspective would examine the informal relationships within the larger bureaucratic organization.
   A. functionalist
   B. interactionist
   C. conflict
   D. none of the above

18. Which of the following is NOT one of the four basic principles of McDonaldizaton?
   A. Predictability
   B. Efficiency
   C. Calculability
   D. Originality

19. Which of the following is NOT an example of what the organization featured in the video is attempting to do differently to offset the limitations of bureaucracies?
   A. Cross-training employees
   B. Using decentralized teams in stores
   C. Using participative decision making
   D. Limiting access to upper management

20. Encouraging employees to do everything they can to please the customer is one way that the organization featured in the video is attempting to offset the limitation of bureaucratic _____.
   A. alienation
   B. anomie
   C. isolation
   D. none of the above

Essay Question
Answer the following question using complete sentences in a well-developed essay.

1. List and explain three advantages and three disadvantages of working within a large bureaucracy.

ANSWER KEY

The following provides the answers and references for the practice test questions.

Multiple Choice:
1. B ....LO 1 ....textbook, p. 143
2. A ....LO 1 ....textbook, p. 143
3. B ....LO 1 ....textbook, p. 143
4. B ....LO 1 ....textbook, p. 143; video segment 2
5. B ....LO 1 .... video segment 2
6. B ....LO 1 .... video segment 2
7. A ....LO 2 ....textbook, p. 143
8. B ....LO 2 ....textbook, pp. 143–144
9. B ....LO 2 ....textbook, p. 143
10. A ....LO 2 ....textbook, pp. 143–144
11. C ....LO 2 ....textbook, pp. 143–144; video segment 2
12. C ....LO 2 ....textbook, p. 143; video segment 2
13. A ....LO 2 ....textbook, p. 145
14. D ....LO 2 ....textbook, p. 145
15. C ....LO 3 .... video segment 4
16. D ....LO 3 .... video segment 4
17. B ....LO 3 .... video segment 4
18. D ....LO 4 .... video segment 5
19. D ....LO 5 .... video segment 6
20. A ....LO 5 .... video segment 6

Essay Question
Answers should include the following types of statements/points:

1. ....LO 2, 3 .... textbook, pp. 143–148; video segments 2 and 6

Advantages:
- Job responsibilities clear
- Clear supervisory structure
- Rules are written down
- Job performance evaluated
- Career Ladder posted so all have opportunity
- Increased productivity can result in pay increase

Disadvantages:
- Can "limit" an employee
- Employee isolated from upper-level supervisors
- Rules only deal with majority; no system for special circumstances
- Often personal likes/dislikes affect evaluation
- Can necessitate "going through the process" even when qualified to perform job
- Increased productivity may lead to layoffs, downsizing

LESSON CONTRIBUTORS

Juanita Firestone, Professor of Sociology, University of Texas at San Antonio, San Antonio, TX

Roger E. Herman, Chief Executive Officer, The Herman Group, Greensboro, NC

Philip Luck, Assistant Professor of Sociology, Lyndon State College, Lyndonville, VT

Parker J. Palmer, Author and Consultant, Madison, WI

George Ritzer, Professor of Sociology, University of Maryland, College Park, MD

Dale E. Yeatts, Chair of Sociology Department, University of North Texas, Denton, TX

Special acknowledgement and thanks go to:
>*James Roe*, Metro Marketing Director, Whole Foods Market
>*Rawland Watson, Jr.*, Area Supervisor, Whole Foods Market
>*Whole Foods Market*, Dallas, Texas
>*United States Naval Academy*, Annapolis, Maryland

Lesson 9

Communities, Societies, and Nations

OVERVIEW

As a political entity, America is a nation. Communities, where people gain a sense of identity and belonging, comprise this nation. America is also a society, with complex social structures organized to meet the needs of its people. Within this nation, this society, and these communities, individuals live, work, and play together in groups, creating the meaning and values that shape their lives. American society is more than just groups of people. It is a meaningful social structure that organizes and bonds its members to carry out the major functions of life, such as reproduction, sustenance, shelter, and defense.

In this video lesson, you will learn that America, like all societies, is not static. You will learn about different types of societies and how they change at varying rates of speed. You will explore societal change in terms of the characteristics of societies: *gemeinschaft* and *gesellschaft*, *primary* and *secondary* groups, and *role* and *status*.

You will learn more about what it means to be a nation through Chad Smith, Principal Chief of the Cherokee Nation. The Cherokee Nation has been a domestic nation since it was designated as such by a U.S. Supreme Court ruling in the case of *Worcester v. Georgia* and the *Cherokee Nation v. Georgia* in 1830 and 1832.

In this video lesson, you will hear from community members about types of communities within the city of Los Angeles. One is considered a territorial community and another is considered a non-territorial community. The virtual community is a third type of community. What are the differences and similarities in communities in terms of how members interact?

You will travel by video to Santa Fe, New Mexico, an exclusive arts mecca and a destination for wealthy Asians, Europeans, and Hollywood celebrities. While visiting Santa Fe, you will explore why change is inevitable when newcomers enter a community and how the reactions to this change vary. The debate over change in Santa Fe has been going on for a long time. The challenges faced by Santa Fe are typical of many communities in today's mobile society. How might the manner in which we address those challenges determine the future of our communities?

LESSON ASSIGNMENTS

Text: William Kornblum: *Sociology in a Changing World*, Chapter 4, "Societies and Nations," pp. 79–93; Chapter 6, "Interaction in Groups," (Communities) pp. 130–132; Chapter 19, "Politics and Political Institutions," pp. 500–506; and Chapter 21, "Population, Urbanization and, the Environment" p. 574 (Private Communities)

Video: "Communities, Societies and Nations" from the series *Exploring Society, Introduction to Sociology*

LESSON GOAL

After completing this lesson, you will be able to discuss the interdependence among, and within, communities, societies, and nations.

LESSON LEARNING OBJECTIVES

1. Describe the different types of societies, noting the characteristics of *gemeinschaft* and *gesellschaft*, *primary* and *secondary groups*, and role and status within each type.
2. Explain the characteristics of a nation.
3. Describe the characteristics and types of communities that exist in society.
4. Describe how change can affect communities.

REVIEW

The following steps are intended to help you learn the material in this lesson. To maximize your learning experience:

a. Scan the Lesson Focus Point questions.
b. Read the assigned text pages.
c. View the video.
d. Write answers to the Lesson Focus Point questions.
e. Complete the Related Activities assigned by your instructor. If none are assigned, use them to help you review the lesson material.
f. Take the Practice Test and check your answers with the Answer Key located at the end of the lesson.

LESSON FOCUS POINTS

1. When did hunting-and-gathering societies begin to develop permanent settlements? What is the significance of this?
2. What social changes made human life take the form it does today?
3. When did human societies develop language?
4. What are the characteristics of the hunting-and-gathering society?
5. What are the characteristics of the pastoral society? Horticultural society?
6. Why was the domestication of animals important?
7. What are the most important factors of social evolution for the agrarian society?
8. When did social stratification develop in society?
9. What is the social order related to the Industrial Revolution?
10. What changes took place in the social structure of American society that encouraged the development of an industrial society?
11. What is the theory of post-industrial society?
12. What are the different types of societies? What are their characteristics?
13. What are *gemeinschaft* and *gesellschaft*?
14. What are primary and secondary groups?
15. What is *status*? What is *master status*? Give examples.
16. What is *role, role conflict*, and *role strain*? Give examples.
17. How do status and role differ in each of the different types of societies?
18. What is the difference between a state and a nation?
19. Is Nigeria a nation or a state?
20. What are *communities*?
21. What is the difference between territorial/nonterritorial communities?
22. What functions do communities serve?
23. What are *electronic communities*?
24. How and why do communities change?
25. What do we mean by a "sense of community"?
26. What happens when newcomers "invade" a community?
27. Why are "gated communities" popular?

RELATED ACTIVITIES

1. Describe at least two examples of ways in which residents of your neighborhood or other neighborhoods in your city try to "defend" the neighborhood from "invasion."

2. Based on your experience or that of someone you know who has lived in or knows a lot about a small town, describe the advantages and the disadvantages of community in that town. Describe specific instances or situations to illustrate your points.

3. Describe at least one non-territorial community to which you or someone you know belongs. What are the personal and social benefits of belonging to this community?

4. What are three advantages and three disadvantages of living where you do? Name the town or city, and describe the advantages and disadvantages you have identified.

PRACTICE TEST

Multiple Choice
Choose the BEST answer.

1. Which of the following changes enabled human life to take the form it does today?
   A. Development of upright posture
   B. Social control of sexuality
   C. Establishment of band of hunter-gatherers as the basic territorial unit of human society
   D. All of the above

2. From the standpoint of social evolution, which of the following is NOT a dimension of agrarian societies?
   A. Escape from dependence on food sources over which humans have no control
   B. Requirement of a supply of land
   C. Need to store and defend food surpluses
   D. Need to wander over lands

3. _____ is the process whereby members of society are sorted into different statuses and classes based on differences in wealth, power, and prestige.
   A. Social stratification
   B. Social leveling
   C. Class division
   D. Class differentiation

4. The _____ marked a dramatic shift from agrarian societies to industrial societies.
   A. Civil War
   B. World War II
   C. Industrial Revolution
   D. growth of cities

5. Which of the following would be an example of *gemeinschaft* interactions within a post-industrial society?
   A. July 4
   B. Presidents' Day
   C. Christmas
   D. September 11, 2001

6. Which of the following apply to a post-industrial society?
   A. Blurring of traditional statuses for men and women
   B. Blurring of traditional statuses regarding age
   C. Changing in terms of roles attached to specific statuses
   D. All of the above

7. In an agrarian society, you saw the emergence of the _____.
   A. primary group
   B. secondary group
   C. interactive
   D. cooperative

8. The _____ is a society's set of political institutions.
   A. nation
   B. state
   C. politics
   D. none of the above

9. Which of the following is NOT a characteristic of a nation?
    A. Sovereignty
    B. Legitimacy
    C. Boundaries
    D. None of the above

10. In a country as large as Africa, one of the fears people have is that if the tribes incorporate and become a modern nation, _____.
    A. the tribes will lose their identities
    B. tribal kings will lose power
    C. tribal rights will be lost
    D. all of the above

11. _____ indicates the right to govern.
    A. Legitimacy
    B. Sovereignty
    C. Righteousness
    D. Politic

12. Which of the following is an example of an African nation?
    A. Ibo
    B. Junkun
    C. Nigeria
    D. Yoruba

13. _____ are sets of primary and secondary groups in which the individual carries out important life functions.
    A. Neighborhoods
    B. Societies
    C. Associations
    D. Communities

14. When people speak of a professional community, such as the medical or legal community, they are referring to a _____ community.
    A. territorial
    B. neighborhood
    C. nonterritorial
    D. vocational

15. More than a nation of geographic boundaries, the Cherokee Nation is a nation of _____.
    A. history
    B. ceremony
    C. people
    D. all of the above

16. Sherman Oaks is an example of a _____ community.
    A. geographic
    B. slated
    C. territorial
    D. planned

17. Which of the following is true about communities?
    A. Something that we feel
    B. Defined by places
    C. Defined by associations
    D. All of the above

18. The Internet has made it possible for _____ communities to form all over the world.
    A. territorial
    B. global
    C. non-territorial
    D. interest

19. An example of how change has affected Santa Fe in both positive and negative ways is _____.
    A. integrated neighborhoods
    B. increased education
    C. increased diversity in the local population
    D. outsiders bringing money and new businesses to the community

20. _____ are becoming a part of Santa Fe neighborhoods.
    A. Gated communities
    B. Integrated neighborhoods
    C. Rental houses
    D. All of the above

Essay Questions

Answer the following questions using complete sentences in a well-developed essay.

1. Define and give an example of role conflict and role strain.

2. Describe and give an example of a gated community. What are three advantages of living in a gated community and the disadvantages of living in such a community?

ANSWER KEY

The following provides the answers and references for the practice test questions.

Multiple Choice:

1. D......LO 1 ....textbook, p. 80
2. D......LO 1 ....textbook, pp. 81–82
3. A......LO 1 ....textbook, pp. 82–83
4. C......LO 1 ....textbook, pp. 85–86; video segment 2
5. D......LO 1 ....textbook, p. 89; video segment 2
6. D......LO 1 ....textbook, pp. 86 and 89; video segment 2
7. B......LO 1 ....video segment 2
8. B......LO 2 ....textbook, p. 91
9. D......LO 2 ....textbook, pp. 91–93
10. D......LO 2 ....textbook, pp. 91–93
11. B......LO 2 ....video segment 3
12. C......LO 2 ....textbook, p. 92
13. D......LO 2 ....textbook, pp. 130–131
14. C......LO 2 ....textbook, p. 131
15. C......LO 2 ....video segment 3
16. C......LO 3 ....textbook, p. 131; video segment 4
17. D......LO 3 ....textbook, pp. 130–132; video segment 4
18. C......LO 3 ....video segment 4
19. D......LO 3 ....video segment 4
20. A......LO 3 ....video segment 4

Essay Questions

Answers should include the following types of statements/points:

1. ....LO 1...textbook, p. 90

   Role conflict:
   When one social role conflicts with another social role.
   Example: as a student, you need to study for an upcoming test. (social role = student) while at the same time your job requires you to work later. (social role = employee) The roles of student and employee are in conflict.

   Role strain:
   When there are conflicting demands on an existing role.
   Example: as a single mother, you need to be at the PTA meeting for your daughter but you also need to be at your son's piano recital. The role of mother is being strained by dual expectations.

2. ....LO 4...textbook, p. 574; video segment 5

   A gated community is a residential area that is encircled by a fence or barrier of some kind. Only people who live in the community can gain access.
   Advantages:
   - Security
   - Property value remains stable
   - Homogenous community sharing similar values

   Disadvantages
   - Isolated from reality
   - Lack of diversity
   - Elitism

LESSON CONTRIBUTORS

Edward Archuleta, Director, Santa Fe Project, 1,000 Friends of New Mexico, Santa Fe, New Mexico

Ralph B. Brown, Associate Professor of Sociology, Brigham Young University, Provo, UT

Paul Duran, County Commissioner, Santa Fe, New Mexico

Arlene N. LewAllen, Director, LewAllen Contemporary, Santa Fe, New Mexico

Cris Moore, City Councilor, Santa Fe, New Mexico

Chad Smith, Principal Chief, Cherokee Nation, Tahlequah, Oklahoma

Karl H. Sommer, Sommer, Fox, Udall, Othmer & Hardwick, Santa Fe, New Mexico

Timothy L. Sullivan, Professor of Sociology and Anthoropology, Cedar Valley College, Lancaster, TX

Terry Williams, Associate Professor, New School University, New York, NY

Special acknowledgement and thanks go to:
   Karla Gonzalez, Santa Fe, New Mexico
   Bishop Stewart, Sherman Oaks, California
   Cherokee Nation, Tahlequah, Oklahoma
   TVI Actors Studio, Sherman Oaks, California

Lesson 10

Social Stratification

OVERVIEW

There are people who hold great wealth, power, and status—and those who don't. Societies have been divided into different social levels, or strata. This is called *social stratification*.

When the *Titanic* hit the iceberg on that April night in 1912, passengers' social levels and social statuses contributed to survival. Consider some of the social facts about the survivors:

- More than 60% were from the wealthy, first-class deck.
- About 36% were from the second-class deck.
- Only 24% were from the lowest or "steerage" class.
- In first class, 97% of women and children were saved.
- In second class, 89% of women and children survived.
- Only 42% of women and children in steerage were saved.

For the passengers of the *Titanic*, the social level or strata meant more than just differences in the types of comforts they enjoyed or food they ate. On that April night, it literally meant life or death.

Today, our chances of survival or living well are largely influenced by the social stratification system of our society. In India, the caste stratification system is a rigid system that was officially abolished by the Indian government in 1949. However, its influence still exists, closely tied to religious and cultural beliefs. For Europeans, the estate system historically consisted of three groups—the nobility, the clergy, and the working class. Closely tied to land ownership, this stratification system was also rigid, without opportunity for social mobility. Here in the United States of America, social mobility is most often associated with the class system of social stratification. This system uses specific criteria for determining one's social class—income, education, and occupation.

In this video lesson, sociologists explore social stratification using the sociological perspectives to understand why social stratification is such an enduring element of society. Whether examining the inequality of power, society's need to differentiate society's rewards, or the shifting concepts of prestige, sociologists provide society with insights about social stratification.

In this video lesson, you'll meet sociologist Elaine Bell Kaplan and hear how she achieved upward social mobility. Her journey is not without personal struggle and heartache. Dr. Kaplan grew up in Harlem—one of seven children in a family that struggled each day. She recalls getting on the bus in Harlem and riding to the museums and cultural centers in New York City. These visits opened up the world to her and early in her life, she made the decision to leave Harlem. Education was her way out of Harlem, and it paved the road to her present position at the University of Southern California.

While we claim that America is the "land of opportunity," we also live with the possibility of losing that which we have achieved. The emotional journey that occurs in situations of loss is often a journey of hurt and confusion. Such situations are not as visible as the differences that were apparent on the *Titanic*, where survival was determined largely by the type of ticket you held.

LESSON ASSIGNMENTS

Text: William Kornblum: *Sociology in a Changing World*, Chapter 10, "Stratification and Global Inequality," pp. 233–260

Video: "Social Stratification" from the series *Exploring Society, Introduction to Sociology*

LESSON GOAL

After completing this lesson, you will be able to explain the concept of social stratification.

LESSON LEARNING OBJECTIVES

1. Describe the systems of social stratification.
2. Explain social stratification using the sociological perspectives.
3. Describe and give examples of the types of social mobility.

REVIEW

The following steps are intended to help you learn the material in this lesson. To maximize your learning experience:

a. Scan the Lesson Focus Point questions.
b. Read the assigned text pages.
c. View the video.
d. Write answers to the Lesson Focus Point questions.
e. Complete the Related Activities assigned by your instructor. If none are assigned, use them to help you review the lesson material.
f. Take the Practice Test and check your answers with the Answer Key located at the end of the lesson.

LESSON FOCUS POINTS

1. What is *social stratification*?
2. Describe the social stratification systems *caste*, *estate*, and *class*.
3. Give examples of each of the systems of social stratification.
4. What is *ascribed* status? *Achieved* status? *Status group*?
5. How are life chances related to social stratification?
6. What is *demeanor*? *Deference*?
7. Why does the caste system still exist in India although officially banned?
8. What is *social mobility*?
9. What is *intergenerational mobility*?
10. What is *intragenerational mobility*?
11. What is *structural mobility*?
12. What is an *open* social stratification system? *Closed* social stratification system?
13. Using the functionalist perspective, explain social stratification.
14. Using the conflict perspective, explain social stratification.
15. Using the interactionist perspective, explain social stratification.
16. What societal forces/changes did Marx not envision?
17. Tell the story of Dr. Elaine Bell Kaplan as it relates to social stratification and social mobility.

RELATED ACTIVITIES

1. Pretend that you have won a $2 million lottery. Let us assume that you wish to use the money to become upwardly mobile. How would you spend it? Would your choices be based on your present tastes? If not, how would you acquire "higher" tastes? How would your relationships with friends change? What would be the psychological and social costs of this new life to you and your family?

2. Are you class conscious? Do you accept your social class as right, proper, and equitable? Explain why or why not.

3. Describe an individual in your community who has a lot of prestige or power or both but not much wealth. What social class would you put that person in? Explain how he or she illustrates either Karl Marx's or Max Weber's view of social stratification and mobility.

4. Do you know of a divorced woman who has experienced downward mobility? Describe the evidence of her downward mobility.

5. Give an example of someone you know who used education to attain upward social mobility.

PRACTICE TEST

Multiple Choice
Choose the BEST answer.

1. Which of the following is an attribute that is used for ranking people and distributing rewards?
   A. Power
   B. Prestige
   C. Age
   D. All of the above

2. _____ are social strata into which people are born and in which they remain for life.
   A. Estates
   B. Castes
   C. Classes
   D. Fixed strata

3. *Class* social strata are based primarily on _____ criteria.
   A. inherited
   B. assigned
   C. economic
   D. accomplished

4. In the United States, very rich and prestigious families often form a _____.
   A. social strata
   B. status group
   C. stratus group
   D. social group

5. In the United States, which of the following are two important criteria for entry into the highest levels of upper-class society?
   A. Money; family prestige
   B. Philanthropy; money
   C. Family prestige; education
   D. Family prestige; philanthropy

6. The principal forces that produce stratification are related to the ways in which people _____.
   A. earn their living
   B. educate themselves
   C. are raised
   D. gain their family name

7. The people who are at or near the bottom of a social stratification system _____.
   A. lack wealth
   B. lack opportunities
   C. lack power
   D. all of the above

8. In this video lesson, while _____ percent of the women and children in the *Titanic's* first class were saved, only _____ percent of the women and children in steerage were saved.
   A. 90; 80
   B. 85; 10
   C. 97; 42
   D. 50; 30

9. The ordering of social groups and individuals hierarchically is

   _____.
   A. social ordering
   B. social stratification
   C. social leveling
   D. social place

10. The _____ social stratification system was composed of _____ groups.
    A. caste; five
    B. estate; three
    C. estate; six
    D. caste; eight

11. In this video lesson, which of the following is NOT one of the factors influencing social class movement?
    A. Economic conditions
    B. Education
    C. Personal effort
    D. All of the above are factors

12. *Demeanor* is the way we present ourselves, including _____.
    A. speech
    B. manners
    C. body language
    D. all of the above

13. Legitimate power is called _____.
    A. authority
    B. coercion
    C. legal power
    D. cultural power

14. Sociologists using the _____ perspective would say the unequal distribution of rewards is necessary.
    A. functionalist
    B. conflict
    C. interactionist
    D. control

15. The concept face is related to _____.
    A. acting non-physically
    B. verbal assaults
    C. status differentiation
    D. tattooing

16. _____ mobility refers to one's chances or rising to or falling from one social class to another within one's own lifetime.
    A. Intergenerational
    B. Self
    C. Lifetime
    D. Intragenerational

17. One reason upward social mobility has been easier to achieve in the United States is because _____.
    A. mobility does not occur in other countries
    B. the class system exists in America
    C. mobility does not occur in other stratification systems
    D. education is available

18. In this video lesson, sociologist Elaine Bell Kaplan's personal story is a good example of _____ mobility.
    A. gender
    B. downward
    C. upward
    D. none of the above

19. One of the reasons Dr. Kaplan wanted to leave her neighborhood was _____.
    A. the schools didn't teach
    B. the high use of drugs
    C. there was no work
    D. all of the above

20. Both downward social mobility and upward social mobility is often associated with _____.
   A. a better life
   B. more friends
   C. more personal peace
   D. emotional tolls

Essay Question
Answer the following question using complete sentences in a well-developed essay.

1. Do you know of a divorced woman who has experienced downward mobility? Describe the evidence of her downward mobility.

ANSWER KEY

The following provides the answers and references for the practice test questions.

Multiple Choice:
1. D..............LO 1 ....textbook, p. 234
2. B..............LO 1 ....textbook, p. 235; video segment 2
3. C..............LO 1 ....textbook, p. 235
4. B..............LO 1 ....textbook, p. 235
5. A..............LO 1 ....textbook, pp. 235–236
6. A..............LO 1 ....textbook, p. 236
7. D..............LO 1 ....textbook, p. 240
8. C..............LO 1 .... student course guide overview, pp. 91–92 ; video segment 2
9. B..............LO 1 .... video segment 2
10. B..............LO 1 .... video segment 2
11. D..............LO 1 .... video segment 2
12. D..............LO 2 ....textbook, pp. 241–242
13. A..............LO 2 ....textbook, p. 243
14. A..............LO 2 ....textbook, p. 251; video segment 3
15. C..............LO 2 .... video segment 3
16. D..............LO 3 ....textbook, p. 249; video segment 4
17. B..............LO 3 .... video segment 4
18. C..............LO 3 .... video segment 5
19. B..............LO 3 .... video segment 5
20. D..............LO 3 .... video segment 5

Essay Question

Answers should include the following types of statements/points:

1. ....LO 3...................................... textbook, p. 249 and pp. 257–258; video segment 4

    Answer Explanation:
    - Decreased income/buying power
    - Lack of established credit
    - Loss of home; moving into a "lesser" place
    - Loss of insurance coverage
    - Increasing cost of childrearing (including day-care)
    - Loss of status
    - "Necessary" employment

LESSON CONTRIBUTORS

Elaine Bell Kaplan, Associate Professor of Sociology, University of Southern California, Los Angeles, CA

Michèle Lamont, Professor of Sociology, Princeton University, Princeton, NJ

Tom Mayer, Professor of Sociology, University of Colorado—Boulder, Boulder, CO

Mary Patillo, Associate Professor of Sociology, Northwestern University, Evanston, IL

Nathaniel Eugene Terrell, Chairperson, Sociology and Anthropology, Emporia State University, Emporia, KS

Special acknowledgement and thanks go to:
    The Kaplan Family, Los Angeles, California

Lesson 11

Social Class

OVERVIEW

When most of us think of social class in our everyday lives, we are likely to define class using a combination of observable characteristics such as speech patterns, material possessions, mannerism, and style of dress. Sociologists also consider information such as income, wealth, occupational status, and educational attainment. This is a more systematic way of defining social class rather than relying entirely on outward appearances. Even so, the boundaries between classes tend to be blurred. Some sociologists argue that there aren't any discrete classes with clearly defined boundaries but rather a socioeconomic continuum upon which we place individuals.

In the lesson video, sociologists talk about how they define the concepts of wealth, power, and prestige—powerful aspects of social class that are reinforced in tangible and intangible ways. For example, the kind of car we drive, the type of home we live in, where we go on vacation, the hobbies we enjoy where we buy our clothes—all of these can indicate social class. While any single indicator can be misleading, when a sociologist explores several of the indicators of social class, a general picture begins to form.

There are real differences among the classes. In the lesson video, you will hear about differences in education, income, and occupation—three powerful indicators of social class. As the sociologists talk about the indicators of social class, ask which social class would you put yourself in, and why? How are you making that determination? If you are like most people, you will use indicators that are common to more than one social class. But what are the dominant indicators of education, occupation, and income?

Social class implies more than just economic position. Your social class helps you understand the world and where you fit into it. We often derive a description of self-worth, culture, and quality of life from social class.

It is the sociologist who studies the complexities of social class and how social class is determined. How does social class relate to life chances? By viewing social classes as social categories, sociologists provide insightful observations. In American society for example, the probability for being arrested is higher among working-class and poor people. People in those same social classes are more likely to get convicted in a court of law, go to prison, and even receive the death penalty than are people in the upper social classes. People in the lower social classes are more likely to die as a result of homicide, accidents, or inadequate health care than are people in the upper classes. The death rate for Americans with family incomes of less than $9,000 is three times higher than that of people with family incomes of more than $25,000 a year.

Finally, in the lesson video, you will meet the Guerry family. You will spend time with them and learn how they live their lives. You will be given important information about the Guerry family; use this information to draw your own conclusions about their social class and life chances.

LESSON ASSIGNMENTS

Text: William Kornblum: *Sociology in a Changing World*, Chapter 10, "Stratification and Global Inequality," pp. 242–245, Chapter 11, "Inequalities of Social Class," pp. 262–292

Video: "Social Class" from the series *Exploring Society, Introduction to Sociology*

LESSON GOAL

After completing this lesson, you will be able to explain the concept of social class in the United States.

LESSON LEARNING OBJECTIVES

1. Discuss the ways sociologists measure social inequality.
2. Discuss how wealth, power, and prestige are related to social class.
3. Differentiate among the social classes.
4. Given an example, describe the factors that determine life chances.
5. Differentiate among the types of poverty.

REVIEW

The following steps are intended to help you learn the material in this lesson. To maximize your learning experience:

a. Scan the Lesson Focus Point questions.
b. Read the assigned text pages.
c. View the video.
d. Write answers to the Lesson Focus Point questions.
e. Complete the Related Activities assigned by your instructor. If none are assigned, use them to help you review the lesson material.
f. Take the Practice Test and check your answers with the Answer Key located at the end of the lesson.

LESSON FOCUS POINTS

1. What is *social class*?
2. What is *wealth*? How is wealth different than income?
3. How is income distributed in the United States?
4. What is *power*?
5. Who are the powerful in the United States?
6. What is the *power elite*?
7. What is *prestige*?
8. How is prestige related to jobs?
9. What are the subjective and objective methods of determining social class?
10. What are the levels of social class?
11. What are the characteristics of each level of social class?
12. What are the most commonly used measures of inequality?
13. How do sociologists measure social inequality?
14. What are some of the findings from a sociological examination of social inequality?
15. What is *educational attainment*?
16. What is *occupational prestige*?
17. What is *absolute poverty*?
18. What is *official poverty*?
19. What is *relative poverty*?
20. How do sociologists define life chances?

RELATED ACTIVITIES

1. Describe the divisions that exist in your community between the working class and the upper class. What evidence illustrates the gap in income and lifestyle between the two classes?

2. In which social class do you fit? (You may use any of the descriptions of social classes described in the textbook, but mention which one you are using.) Are there aspects of your lifestyle, income, or job that would place you in two social classes?

3. How is your education contributing to your social-class standing now and in the future? How is it likely to affect your health and other aspects of your "life chances"?

4. Do you know a person in the working class who illustrates some of the changes described in the textbook? Into which of the two major divisions of the American working class does or did the person fit?

5. When you were a child, did your parents or others put pressure on you to perform well in school? If so or if not, how did this affect your behavior, attitude, values, and other responses? How has your education affected your life chances?

PRACTICE TEST

Multiple Choice
Choose the BEST answer.

1. Which of the following is NOT one of the most commonly used measures of inequality?
   A. Wealth
   B. Occupation
   C. Family
   D. Educational attainment

2. By 2001, the average pretax household income for the top 20 percent was almost _____.
   A. $100,000
   B. $170,000
   C. $ 80,000
   D. $200,000

3. Which of the following statements is NOT an accurate measurement of social inequality over a specified time period between the years 1970 and 2000?
   A. There is a widening gap between the rich and the poor in the United States.
   B. There is a lessening gap between the rich and the poor in the United States.
   C. The top twenty percent of United States households are receiving more income than the remainder of all households.
   D. None of the above

4. Which of the following western industrial nations has the largest gap between the rich and poor?
   A. Great Britain
   B. Ireland
   C. Mexico
   D. United States

5. Which of the following statements is an accurate measure of social inequality about the level of educational attainment in the United States (2000)?
   A. 16.5 percent of African Americans complete four years or more of college.
   B. Twenty-six percent of whites complete four years or more of college.
   C. Almost eleven percent of Latinos complete four years or more of college.
   D. All of the above

6. In the United States, there is an increasing demand for educated people; thus, education becomes the primary vehicle for _____.
   A. increased incomes
   B. upward mobility
   C. increased earning power
   D. all of the above

7. Occupational _____ is measured using surveys of how people throughout a society rate different jobs.
    A. prestige
    B. attainment
    C. importance
    D. value

8. An occupation at the top of the prestige ranking in the United States is _____; at the bottom is _____.
    A. lawyer; janitor
    B. physician; rag picker
    C. physician; bellhop
    D. minister; truck driver

9. The _____ of determining social-class membership uses interviews in which respondents give their opinions about their class rankings.
    A. objective method
    B. response method
    C. interview method
    D. subjective method

10. The social class most shaped by education is the _____.
    A. upper class
    B. upper middle class
    C. middle class
    D. working class

11. The lower middle class would usually have what level of education?
    A. Public school
    B. Community college
    C. Four year state college
    D. All of the above

12. The social class that is probably experiencing the most rapid social change is the _____.
    A. upper class
    B. upper middle class
    C. lower middle class
    D. working class

13. Store clerks would be an example of a job typical for the _____.
   A. lower middle class
   B. working class
   C. lower class
   D. none of the above

14. Sociologists study life chances related to _____.
   A. social class
   B. social mobility
   C. life opportunity
   D. individual attainment

15. The lower the social class, the more _____ the work.
   A. beneficial
   B. prestigious
   C. unhealthy
   D. meaningful

16. Tony and Kay Guerry work as _____ in their home.
   A. graphic/web designers
   B. upkeep managers
   C. child care facilitators
   D. none of the above

17. One of the ways the Guerrys can increase their income is to _____.
   A. stop working at home
   B. work for a computer firm
   C. expand their business
   D. none of the above

18. The Guerrys are putting money aside each month for _____.
   A. a new house
   B. a new car
   C. their children's education
   D. all of the above

19. Which of the following annual incomes would approximate a family of four being classified as poor?
    A. $25,000
    B. $10,000
    C. $ 5,000
    D. $17,000

20. _____ poverty makes a comparison within a culture, while _____ poverty compares different cultures (poverty in the United States versus poverty in India).
    A. Real; absolute
    B. Absolute; relative
    C. Relative; absolute
    D. True; real

Essay Questions
Answer the following questions using complete sentences in a well-developed essay.

1. How is your education contributing to your social-class standing now and in the future? How is it likely to affect your health and other aspects of your "life chances"?

2. Has your social-class position constrained or helped you in achieving your educational goals? Explain.

ANSWER KEY

The following provides the answers and references for the practice test questions.

Multiple Choice:

1. C .... LO 1 .... textbook, pp. 263–264
2. A .... LO 1 .... textbook, p. 264
3. B .... LO 1 .... textbook, pp. 264–265
4. D .... LO 1 .... textbook, p. 265
5. D .... LO 1 .... textbook, p. 266
6. D .... LO 2 .... textbook, pp. 266–267
7. A .... LO 2 .... textbook, p. 267
8. B .... LO 2 .... textbook, p. 268
9. D .... LO 3 .... textbook, p. 273
10. B .... LO 3 .... video segment 3
11. D .... LO 3 .... video segment 3
12. D .... LO 3 .... video segment 3
13. B .... LO 3 .... video segment 3
14. A .... LO 4 .... textbook, pp. 274–275; video segment 4
15. C .... LO 4 .... video segment 4
16. A .... LO 4 .... video segment 4
17. C .... LO 4 .... video segment 4
18. C .... LO 4 .... video segment 4
19. D .... LO 5 .... textbook, p. 281
20. B .... LO 5 .... textbook, pp. 281–282

Essay Questions

Answers should include the following types of statements/points:

1. .... LO 2, 3 .... textbook,, pp. 275–284; video segments 2 and 3

    Education provides a means to:
    *   More employment opportunities
    *   Supervisory positions
    *   More money
    *   Jobs with benefits (health coverage, etc.)
    *   Social class mobility
    *   Family health benefits (more and better health care)
    *   Homes located in "nice neighborhoods"
    *   Better schools (more dollars for each student)
    *   Less exposure to drugs, crime
    *   Schools provide more extracurricular activities
    *   Family support; education valued
    *   Better nutrition

---

2.   ....LO 4 ....textbook,, pp. 275–284; video segment 4

Usually, the higher one's social-class position, the more opportunity or access he or she has to better education. An exception would be access through scholarships.

LESSON CONTRIBUTORS

Tracey McKenzie Elliott, Professor of Sociology, Collin County Community
    Colleges, Frisco, TX
Elaine Bell Kaplan, Associate Professor of Sociology, University of Southern
    California, Los Angeles, CA
Tom Mayer, Professor of Sociology, University of Colorado—Boulder,
    Boulder, CO
Michèle Lamont, Professor of Sociology, Princeton University, Princeton, NJ

Special acknowledgement and thanks go to:
    The Guerry Family, Garland, Texas
    Park Place Dealerships, Dallas, Texas

Lesson 12

Gender

OVERVIEW

It's been a long time since Margaret Mead's research (1935) examined gender differences in several world cultures. Mead's contemporaries usually assumed that such differences were related to biological factors. Contrary to popular assumptions, Mead found that gender differences in various cultures were products of socialization. What it meant to be a woman or a man varied tremendously among the cultures that she studied.

But what about gender differences in the United States? People born in the early part of the twentieth century grew up during a time when the accepted understanding of differences between behaviors for men and women was restricted to an explanation defined in terms of gender. The Women's Movement challenged this gender-restricted understanding of behavioral differences between women and men. Even today, sociologists continue to question how children are socialized into specific gender role behaviors and expectations.

Other gender questions that sociologists explore are: How do gender differences influence access to wealth and power in America? Why does society continue to value the work that men do more than the work that women do? And why, after all the attention given to gender differences and inequalities, do women still lag behind men in their salaries—even when they both work in the same occupation?

In this lesson, you will meet Police Chief Barbara Childress. She didn't always find support and acceptance in her job; she knew she had to earn the respect and admiration of her fellow officers. As you learn more about Chief Childress, ask yourself this question: would a male police chief have had the same struggle for acceptance in that job?

You will also meet Marcus White. Marcus loves being a nurse. Sometimes his patients are surprised when he enters their rooms—he is often mistaken as a doctor. Note the types of obstacles that Marcus encounters while performing his duties as a nurse. Do the doctors in the hospital treat him differently than his female coworkers?

Sociology helps us explore the meanings of gender in our society and culture, and offers insights into what it means to be a woman or a man.

LESSON ASSIGNMENTS

Text: Kornblum, *Sociology in a Changing World*, Chapter 5, "Socialization," pp. 120–122, Chapter 13, "Inequalities of Gender," pp. 328–352

Video: "Gender" from the series *Exploring Society, Introduction to Sociology*

LESSON GOAL

After completing this lesson, you will be able to discuss how society's expectations influence the definitions and behaviors associated with gender.

LESSON LEARNING OBJECTIVES

1. Explain the relationship of gender identity to gender socialization.
2. Discuss how gender is related to sexuality.
3. Discuss how fashion relates to gender.
4. Discuss gender inequality.
5. Discuss gender stratification.

REVIEW

The following steps are intended to help you learn the material in this lesson. To maximize your learning experience:

  a. Scan the Lesson Focus Point questions.
  b. Read the assigned text pages.
  c. View the video.
  d. Write answers to the Lesson Focus Point questions.
  e. Complete the Related Activities assigned by your instructor. If none are assigned, use them to help you review the lesson material.
  f. Take the Practice Test and check your answers with the Answer Key located at the end of the lesson.

LESSON FOCUS POINTS

1. Differentiate between sex and gender.
2. What are *gender roles*?
3. What is *sexual harassment*?
4. What is *empowerment*?
5. How does the communications media influence gender?
6. How do agencies of socialization like schools influence gender?
7. What is *gender socialization*?
8. What role does family play in gender socialization?
9. What role does education play in gender socialization?
10. What role does the communications media play in gender socialization?
11. How does the communication media influence the way men and women look at their own bodies?
12. How is fashion related to gender?
13. Historically, what role has fashion played in gender identity?
14. Give some specific examples of fashion trends that signaled changes in a society's concept of gender identity.
15. What is *gender inequality*?
16. Describe the rise of feminism in America.
17. Describe feminism's "second wave."
18. What were the goals of the second wave?
19. How does the institution of education discriminate against females?
20. Describe the goal(s) of the first wave of feminism.
21. What is the "second shift"?
22. Why is the empowerment of women important to sociologists?
23. According to the Labor Department's Bureau of Statistics (2000), how much did women earn compared to every dollar that men earned?
24. What is the glass ceiling?
25. Define the "glass escalator" phenomena and comment on why it exists.
26. What are "pink-collar ghettoes?"
27. What is gender stratification?
28. What is sexism?
29. How has Police Chief Barbara Childress approached gender stratification?
30. What is an example of gender stratification?
31. Identify the percentage of single men and women who live in poverty.
32. What are some misconceptions patients have when meeting Marcus White?
33. What is the *time bind*?

RELATED ACTIVITIES

1. Record the poems, sayings, phrases, and words used to describe little boys and little girls that you have heard from parents, relatives, friends, and teachers throughout your life. List at least three cultural norms about the genders indicated by what you have written.

2. View two television shows in which a woman plays the lead character. List the characteristics of the women based on what you see in the show.

3. Research how the Family Leave Act has affected men by interviewing three men who have used this Act to be with their children.

4. Interview a high school or college administrator to determine which classes offered by that institution are attended predominantly by students of one gender.

5. Interview a man and woman who are at least sixty years of age about how gender role expectations have changed for them over the years.

PRACTICE TEST

Multiple Choice
Choose the BEST answer.

1. _____ is an individual's own feelings about whether she or he sees themselves as a woman or a man.
   A. Gender identity
   B. Self concept
   C. Self awareness
   D. Sexuality

2. In Asia and Africa, norms favor boys over girls—often to the detriment of female development. Which of the following statements reflects the preferential treatment?
   A. Boys receive more food than girls.
   B. Girls often expend more calories on field and house work.
   C. Proteins are often consumed by boys, leaving the leftovers for girls.
   D. All of the above.

3. Which of the following is NOT one of the prevailing beliefs about men and women, according to Sandra Bem?
   A. They have different psychological and sexual natures.
   B. Men are the inherently dominant sex.
   C. Male-female difference and male dominance is natural.
   D. Women display an aggressive nature.

4. Gender socialization begins with the _____.
   A. school
   B. church
   C. hospitals
   D. family

5. One result of young girls trying to look like the women on television is an increase in the number of _____.
   A. eating disorders
   B. tall models
   C. unrealistic images
   D. personality disorders

6. "Sex," in sociological terms, refers to the _____ differences between males and females.
   A. anatomical
   B. physiological
   C. biological
   D. psychosocial

7. One of the forces that had a dramatic impact on women's fashion was _____.
   A. television
   B. radio
   C. World War I
   D. World War II

8. _____ will often use fashion as a way to make commentary about society.
   A. Parents
   B. Social movements
   C. Culture
   D. Schools

9. Sociologists explore gender issues dealing with differential treatment and power by studying gender _____.
   A. access
   B. allotment
   C. inequality
   D. differential

10. What term refers to the process whereby relatively powerless people assert their needs and overcome obstacles to become full participants in society?
    A. Access stratification
    B. Equal opportunity
    C. Full participation
    D. Empowerment

11. The number of people living on less than one dollar a day (globally) constitutes _____ of the world's population.
    A. one-fifth
    B. one-third
    C. one-tenth
    D. one-twelfth

12. Which of the following best illustrates the impact of social structure—as opposed to socialization—on the inequalities and differences between the roles of men and women?
    A. Programming girls that they should be good at domestic roles
    B. Segregating games for boys and girls
    C. Implementing school rules that permit coeducational sports
    D. Teaching cultural norms about feminine and masculine behaviors

13. In the United States, _____ percent of all single women are poor compared to _____ percent of single men.
    A. 40; 10
    B. 36.4; 22.5
    C. 14; 20
    D. 60.5; 30.8

14. In the United States, the gender poverty gap between men and women is primarily the result of _____.
    A. disparities in employment
    B. socialization
    C. institutional sexism
    D. gender bias

15. The objectification of women in the role of the beauty contestant is an example of _____.
    A. sexism
    B. sex ratio
    C. sexual identity
    D. gender validity

16. The first wave of feminism worked not only on the issues of women but also _____.
    A. slavery
    B. children
    C. the Constitution
    D. men

17. The _____ is an invisible barrier that women often encounter in the working environment.
    A. glass dome
    B. steel door
    C. trap door
    D. glass ceiling

18. Gender _____ is a hierarchical valuing of people on the basis of sex.
    A. priority
    B. pyramid
    C. stratification
    D. order

19. When Chief of Police Barbara Childress first applied for a job at the police department she found the only job open to her was police _____.
    A. driver for the chief of police
    B. dispatcher
    C. officer
    D. investigator

20. Marcus White is a _____.
    A. police chief
    B. nurse
    C. doctor
    D. physical therapist

Essay Question
Answer the following question using complete sentences in a well-developed essay.

1. In your family, did your parents assign and perform tasks and responsibilities based on the expectations of gender roles? Give specific examples of the tasks each gender and parent was assigned.

ANSWER KEY

The following provides the answers and references for the practice test questions.

Multiple Choice:
1.   A.............LO 1..textbook, p. 120
2.   D.............LO 1..textbook, p. 121
3.   D.............LO 1..textbook, p. 333
4.   D.............LO 1..video segment 2
5.   A.............LO 1..video segment 2
6.   C.............LO 2..textbook, p. 120
7.   D.............LO 3..video segment 3
8.   B.............LO 3..video segment 3
9.   C.............LO 4......................................textbook, p. 329; video segment 4
10. D.............LO 4..textbook, p. 329
11. A.............LO 4..textbook, p. 329
12. B.............LO 4, 5..textbook, p. 339
13. B.............LO 4..textbook, p. 340
14. A.............LO 4..textbook, p. 340
15. A.............LO 4, 5..textbook, pp. 340–341
16. A.............LO 4..video segment 4
17. D.............LO 4..video segment 4
18. C.............LO 5......................................textbook, p. 336; video segment 5
19. B.............LO 5..video segment 5
20. B.............LO 5..video segment 5

Essay Question
Answers should include the following types of statements/points:

1.   ....LO 1, 5..textbook, pp. 120–121 and pp. 336–339

Tasks	Gender
Helping with housework	Female: daughters, mother
Dishes	Female: daughter, mother
Yard work help	Male: son, father
Repairs in home	Male: father
Grocery shopping	Female: mother
Cooking	Female: mother
BBQ	Male: father
Driving kids to places	Female: mother
Taking kids to doctor's appointment	Female: mother
Taking out the trash	Male: father, son

LESSON CONTRIBUTORS

Juan Battle, Professor of Sociology, Hunter College, New York, NY

Barbara Childress, Chief of Police, Richland Hills Police Department, Richland Hills, TX

Ray Eve, Professor of Sociology, University of Texas—Arlington, Arlington, TX

Arlie Russell Hochschild, Professor of Sociology, University of California—Berkeley, Berkeley, CA

Claire Renzetti, Professor and Chair of Sociology, St. Joseph's University, Philadelphia, PA

Bette Woody, Professor of Sociology, University of Massachusetts, Boston, MA

Marcus Ray White, Registered Nurse, Hurst-Euless-Bedford Hospital, TX

Special acknowledgement and thanks go to:

Campbell Agency, Dallas, TX

Clubhouse for "Kids" Only!!!, Bedford, TX

Enchanted Forest Bookstore, Dallas, TX

Angela Auzenne and family, Dallas, TX

Freddie Castillo and family, Dallas, TX

Dottie Clark and family, Dallas, TX

The Comber Family, Dallas, TX

The Klein Family, Dallas, TX

The Vasquez Family, Dallas, TX

Rick Walker and family, Dallas, TX

The White Family, Flower Mound, TX

Evelyn Wong and family, Dallas, TX

Lesson 13

Race and Ethnicity

OVERVIEW

What is race? What is ethnicity? How are they different? It is the sociologist who explores these concepts, which continue to affect our lives and have an impact on our society.

Historically, sociologists and biologists have defined race differently. Biologists have primarily used physical characteristics such as skin color, hair and bone structure to differentiate between races. Sociologists view race as a social concept that may vary from one society to another, depending on how the people of a society perceive physical differences among human beings.

Sociologist Kay Coder experiences how the *social* concept of race is dependent on where she goes in the world. In Japan, people think of her as white or Caucasian; in England, people think of her as Hispanic; in the United States, people are confused, but know she is not fully Caucasian-American. Professor Coder is an example of how one person, with the same physical characteristics, can be perceived differently depending on where she is in the world. This is what sociologists mean when defining race as a *social* concept.

Ethnicity is different. Ethnicity is based on shared cultural identity. Ethnicity is the choice of the individual. You'll learn about the difference between race and ethnicity through the Cajuns of southern Louisiana, who share a sense of cultural identity—whether in their dance, music, or food.

What can we learn about racial and ethnic differences using the sociological perspectives? How do these perspectives help us better understand these important sociological concepts?

What happens when different groups come together? The patterns of intergroup relations range from intolerance to acceptance and, ideally, toward a celebration of our diversity. You'll hear the stories of people whose lives continue to be affected by how they have been treated based on peoples' perceptions of race and ethnicity. It is through these stories that you can begin to understand the powerful meanings behind race and ethnicity.

LESSON ASSIGNMENTS

Text: William Kornblum: *Sociology in a Changing World*, Chapter 12, "Inequalities of Race and Ethnicity," pp. 294–326

Video: "Race and Ethnicity" from the series *Exploring Society, Introduction to Sociology*

LESSON GOAL

After completing this lesson, you will be able to explain how race and ethnicity influence social patterns of human interaction.

LESSON LEARNING OBJECTIVES

1. Discuss race and ethnicity as sociological concepts.
2. Apply the sociological perspectives to the concepts of race and ethnicity.
3. Describe and give examples of intergroup relations.

REVIEW

The following steps are intended to help you learn the material in this lesson. To maximize your learning experience:

    a. Scan the Lesson Focus Point questions.
    b. Read the assigned text pages.
    c. View the video.
    d. Write answers to the Lesson Focus Point questions.
    e. Complete the Related Activities assigned by your instructor. If none are assigned, use them to help you review the lesson material.
    f. Take the Practice Test and check your answers with the Answer Key located at the end of the lesson.

LESSON FOCUS POINTS

1. Define *race*.
2. Define *ethnicity*.
3. Differentiate between race and ethnicity.
4. What is racism?
5. What is *institutional racism*?
6. What is *prejudice*?
7. What is a *stereotype*?
8. Differentiate between *de jure* and *de facto* segregation.
9. What is *discrimination*?
10. What is *scapegoating*? *Projection*?
11. What is a *racial group, ethnic group, minority group, dominant group*?
12. Other than a group's numerical size, what determines a minority group?
13. Give an example of race defined as a social concept.
14. What is *reverse* discrimination? Affirmative action?
15. Explain race and ethnicity, using the sociological perspectives.
16. Explain the role of prejudice in Nazi Germany.
17. Explain how using the conflict perspective would help us understand the manipulation of ethnic and racial groups during workers' strikes.
18. Explain the significance of racial and ethnic labels.
19. Describe and give examples of patterns of intergroup relations—genocide; expulsion; slavery; segregation; assimilation.
20. What is *Anglo-conformity*?
21. What is the *melting pot*?
22. What is *cultural pluralism*?

RELATED ACTIVITIES

1. Citing the elements of the definition of a minority group given in the textbook as your reference, explain why you are or are not a member of a minority group.

2. Explain the difference between genocide and expulsion. Cite at least one example of each to make your point.

3. What is your racial, ethnic, or national heritage? Would you say you have been assimilated or not? Explain, giving examples or evidence of your assimilation or non-assimilation.

4. Examine your own prejudices. Describe an incident in which you internally pre-judged someone who was of a racial, ethnic, religious, or national group different from your own. What were your judgments and feelings? Do not justify your judgments.

5. Interview a Latino person about his or her culture. Ask questions that will give you insight into that culture: questions about customs, values, family, beliefs, mutual support, art, and other special things about the culture. One of the goals of your interview is to discover elements of that culture upon which society could capitalize. Describe these beneficial aspects of the culture, and explain how you think our society as a whole could benefit.

6. If you have experienced prejudice or discrimination yourself, answer the questions that follow. If you have not, interview someone who has. Write out the responses to the following:
    A. Describe an instance in which you experienced prejudice or discrimination.
    B. What was said and done that showed the prejudice or discrimination?
    C. How did this make you feel?
    D. How did you react visibly?

PRACTICE TEST

Multiple Choice
Choose the BEST answer.

1. Ethnicity is based on _____.
    A. shared cultural identity
    B. shared physical characteristics
    C. shared belief systems
    D. shared DNA

2. Saying, "All women gossip," is an example of a _____.
    A. discrimination
    B. prejudice
    C. stereotype
    D. fad

3. A _____ is an attitude that prejudges a person.
   A. prejudice
   B. stereotype
   C. discrimination
   D. racism

4. The statement, "Some members of the majority think that they are victims of 'reverse discrimination' in which they are penalized for the wrongs of earlier generations," is cited as an argument _____.
   A. in favor of institutional discrimination
   B. against cultural pluralism
   C. in favor of ethnic nationalism
   D. against affirmative action

5. Which of the following characteristics has NOT been used by biologists to describe race?
   A. Skin color
   B. Bone structure
   C. Hair structure
   D. Family background

6. Sociologist Kay Coder talks about how she is treated differently depending on _____.
   A. people's perceptions
   B. physical characteristics
   C. how she acts
   D. her family background

7. Sociologist Theresa Martinez talks about black and white students being given standardized tests with at least equal results for the black students; yet, the black students were being put in the lowest academic tracks. This is an example of _____.
   A. official discrimination
   B. discriminatory practices
   C. official racism
   D. institutional racism

8. When sociologist Kay Coder travels to Japan, she is often perceived as
   _____.
   A. Caucasian
   B. Japanese
   C. American Indian
   D. Hispanic

9. Labeling is one aspect of race and ethnicity that sociologists use the
   _____ perspective to examine.
   A. conflict
   B. interactionist
   C. functionalist
   D. feminist

10. _____ is the process whereby we attribute to other people behaviors and
    feelings that we are unwilling to accept ourselves.
    A. Projection
    B. Scapegoating
    C. Escapism
    D. Assimilation

11. The _____ perspective looks at race and ethnicity as serving important
    purposes in society.
    A. conflict
    B. interactionist
    C. functionalist
    D. feminist

12. The _____ perspective examines race and ethnicity in terms of economic
    and political power.
    A. functionalist
    B. conflict
    C. interactionist
    D. sociological

13. Using the different groups of people as scapegoats was an important part in creating a sense of community for the German people during the Nazi regime in WWII. The sociological perspective that sociologists would use to study this phenomenon is _____.
    A. interactionism
    B. creativism
    C. functionalism
    D. conflict

14. Sociologist Troy Duster makes the point, "Whites are only white if there are non-whites around." Thus, the label "white" derives its meaning only in relationship to something else. The sociological perspective that sociologists would use to study this concept is the _____ perspective.
    A. functionalist
    B. conflict
    C. interactionist
    D. feminist

15. _____ is the forcible removal of a population from a territory.
    A. Genocide
    B. Expulsion
    C. Explosion
    D. Settlement

16. _____ is the ownership of one group by another.
    A. Segregation
    B. Assimilation
    C. Slavery
    D. Integration

17. What type of segregation results from laws or other norms that force one people to be separate from others?
    A. *De jure*
    B. *De facto*
    C. *Ipso facto*
    D. *Ex cathedra*

18. _____ is the legally sanctioned segregation that was law in South Africa until 1991.
    A. Slavery
    B. Apartheid
    C. Integration
    D. Separation

19. A pluralistic society is one in which different ethnic and racial groups are _____.
    A. assimilated into the dominant culture, which then becomes the largest culture
    B. encouraged to separate into their own territories
    C. merged into one melting pot
    D. able to maintain their own culture as they gain equality in the institutions of the larger society

20. According to Vicky Maharaj, there were four recognized groups of people in South Africa. Which of the following does NOT belong?
    A. Whites
    B. Coloreds
    C. Indians
    D. Browns

Essay Questions

Answer the following questions using complete sentences in a well-developed essay.

1. Define and give an example of prejudice and discrimination.

2. List and define the patterns of intergroup relations beginning with those patterns most intolerant to those patterns that are more tolerant.

ANSWER KEY

The following provides the answers and references for the practice test questions.

Multiple Choice:

1. A...............LO 1...textbook, pp. 297–299; video segment 2
2. C...............LO 1...textbook, p. 313; video segment 2
3. A...............LO 1...textbook, p. 313; video segment 2
4. D...............LO 1...textbook, pp. 315–318
5. D...............LO 1...video segment 2
6. A...............LO 1...video segment 2
7. D...............LO 1...video segment 2
8. A...............LO 1...video segment 2
9. B...............LO 2...video segment 3
10. A...............LO 2..textbook, p. 319
11. C...............LO 2..textbook, pp. 319–320; video segment 3
12. B...............LO 2...textbook, p. 320; video segment 3
13. C...............LO 2..textbook, pp. 319–320; video segment 3
14. C...............LO 2...video segment 3
15. B...............LO 3..textbook, pp. 302–304; video segment 5
16. C...............LO 3..textbook, pp. 304–305; video segment 5
17. A...............LO 3..textbook, p. 305
18. B...............LO 3..textbook, pp. 305–306; video segment 5
19. D...............LO 3...textbook, p. 309 and p. 312; video segment 5
20. D...............LO 3...video segment 5

Essay Questions

Answers should include the following types of statements/points:

1. ....LO 1...Kornblum, pp. 313–314; video segment 2
    Definitions/examples:
    - Prejudice – an attitude that prejudges a person based on the group that person is a member of. Example: All women gossip.
    - Discrimination – actual unfair treatment of people based on the group membership. Example: Refusing to rent an apartment to someone because that person is Latino.

2. ....LO 3...Kornblum, pp. 301–313; video segment 5
    Patterns of intergroup relations:
    - Genocide – intentional extermination of a population
    - Expulsion – forcible removal of a population from a territory
    - Slavery – ownership of one population by another
    - Segregation – separation (legal) of groups of people (race or ethnicity)
    - Assimilation – group blends into the majority population

LESSON CONTRIBUTORS

Juan Battle, Professor of Sociology, Hunter College, New York, NY
Kay Coder, Professor of Sociology, Richland College, Dallas, TX
Troy Duster, Professor of Sociology, New York University, New York, NY
Carl Knight, Professor of Anatomy, Eastfield College, Mesquite, TX
Theresa A. Martinez, Associate Professor of Sociology, University of Utah, Salt
    Lake City, UT

Special acknowledgement and thanks go to:
    Esat, Kujtim, and *Vlora Gllareva*, Dallas, TX
    Sofia Kusbri, Dallas, TX
    Edward Hummingbird, Director, Blue Deer Gallery, Dallas, TX
    Michael Jacobs, Dallas Memorial Center for Holocaust Studies, Dallas, TX
    Vicky P. Maharaj, Press Secretary, Embassy of South Africa, Washington,
      DC
    Donald Payton, Historian, Dallas, TX
    Miller Family Reunion, Dallas, TX
    Dallas Memorial Center for Holocaust Studies, Dallas, TX
    Randol's Restaurant, Lafayette, LA
    Lee Benoit & The Bayou Stompers, Rayne, LA
    Embassy of South Africa, Washington, D.C.

Lesson 14

Age

OVERVIEW

Age determines a great deal about the opportunities open to us and the roles that we play. We go through our lives with those who were born about the same time as us—our age cohort. We celebrate rites of passage together, adding significance to the social meanings of age.

Due to increasing life expectancy, the composition of America's population is shifting. We're getting grayer in larger numbers than ever before. American life expectancy at birth in 1900 was about forty-nine years. Today, according to the Bureau of Census, male children born in 1995 are projected to live an average of seventy-two years and female children born in 1995 are projected to live an average of seventy-nine years. These life expectancies are projected to increase to eighty-six years and ninety-two years respectively by the year 2050. Americans are also having fewer children. This shifting birth rate combined with a declining death rate is creating a larger older population proportionally. In 1920, 4.6 percent of the U.S. population was aged sixty-five or older. In 1984, this figure reached 11.8 percent. By 2030, the projection is that at least one in five Americans will be elderly.

What are the global and social implications of an aging population? What opportunities are available to us as we age? How has America's aging population changed our perceptions of what it means to age? How can sociology help us understand how age is socially constructed in our society?

In this lesson, you'll meet sociologists who specialize in gerontology, a discipline focusing on the study of age. You will hear them talk about the impact aging has on society, including important challenges that America faces as her citizens age.

LESSON ASSIGNMENTS

Text: William Kornblum: *Sociology in a Changing World*, Chapter 14, "Inequalities of Youth and Age," pp. 354–376

Video: "Age" from the series *Exploring Society, Introduction to Sociology*

LESSON GOAL

After completing this lesson, you will be able to discuss the sociological definition of age and the social implications of an aging population.

LESSON LEARNING OBJECTIVES

1. Discuss age in sociological terms.
2. Discuss aging from each of the sociological perspectives.
3. Examine the social implications of an aging population.
4. Explain how patterns of inequality influence age.

REVIEW

The following steps are intended to help you learn the material in this lesson. To maximize your learning experience:

a. Scan the Lesson Focus Point questions.
b. Read the assigned text pages.
c. View the video.
d. Write answers to the Lesson Focus Point questions.
e. Complete the Related Activities assigned by your instructor. If none are assigned, use them to help you review the lesson material.
f. Take the Practice Test and check your answers with the Answer Key located at the end of the lesson.

LESSON FOCUS POINTS

1. Why are sociologists interested in age?
2. How has an aging population influenced society?
3. Define and give examples when appropriate of the following terms: *life expectancy*, *age cohort*, *rites of passage*, *gerontology*, *age grade*, and *life course*.
4. What is the *baby boom*?
5. What is the *baby boom echo*? *Baby boomlet*?

6. What does the *grouping of populations* mean?
7. What are the stages of aging?
8. What is the functionalist perspective of age?
9. What is the conflict perspective of age?
10. What is the interactionist perspective of age?
11. What is the *baby-boom age cohort*?
12. What does the expression "pig in a python" refer to?
13. What is *ageism*?
14. What is the significance of the name change for AARP?
15. How do labels and stereotypes related to age affect society?
16. How have movies depicted age? Give an example.
17. What are the social implications of an aging population?
18. What are some programs in which older persons volunteer?
19. What is *structural lag*?
20. How are older persons remaining active contributors to society?
21. How has care-giving been redefined?
22. In the video lesson, why does Ted Pohrte continue to work?
23. How is aging portrayed in films?
24. How does age inequality affect life chances?
25. How is age inequality affected by race? Gender?
26. What is "The Challenge of Youth"?
27. How are youth subcultures related to youth peer groups?
28. Describe some of the variations in the value placed on children?
29. What are some indicators of the well-being of American children?
30. What are examples of stereotypes of the elderly?
31. What is the *disengagement theory*? *Activity theory*?
32. What are some examples of social movements among the elderly?
33. What is the *hospice movement*?

RELATED ACTIVITIES

1. Describe a rite of passage you have gone through. What were your feelings about yourself before and after? How did people treat you differently after the experience?

2. Record comments you hear during the next week or two about older or younger people or the process of aging. What sorts of images of the aging process emerge from these comments? Give examples and summarize your findings.

3. Describe three common stereotypes of older or younger people. Why do such stereotypes exist?

4. Find five signs that are age-related (example: must be 18 to purchase.......). Take a picture of each sign and then write an explanation that supports the age restriction depicted in the sign.

PRACTICE TEST

Multiple Choice
Choose the BEST answer.

1. Which of the following is NOT an example of an age strata?
   A. Teenagers
   B. Weddings
   C. Preschoolers
   D. Infants

2. Demographers use age _____ in studying how populations change.
   A. grades
   B. strata
   C. groups
   D. cohorts

3. The *baby boomlet* or the *baby boom echo* refers to _____.
   A. children of parents who were part of the original baby boom
   B. the baby boom generation
   C. the parents of baby boomers
   D. none of the above

4. A bar mitzvah is an example of _____.
   A. an age strata
   B. an age grade
   C. a rite of passage
   D. an age cohort

5. People born between the years 1950 and 1954 would be members of an _____.
    A. age cohort
    B. age level
    C. age stage
    D. none of the above

6. The average number of years a member of a given population can expect to live beyond his or her present age is called _____.
    A. life chances
    B. life expectancy
    C. life strata
    D. age cohort

7. Which of the following has NOT had an impact on life expectancy?
    A. Technology
    B. Medicine
    C. Number of births
    D. Nutrition

8. Which of the following is a *social* effect of age?
    A. Age minimum for a driver's license
    B. Age requirement for entering school
    C. Age minimum to receive a senior discount
    D. All of the above

9. Which of the following would a sociologist be interested in studying as it relates to the social impact of age?
    A. Education
    B. Housing
    C. Crime
    D. All of the above

10. The _____ perspective would look at how age relates to particular roles in society.
    A. functionalist
    B. conflict
    C. interactionist
    D. feminist

11. The idea that older workers would retire and be replaced with younger workers would *make sense* in the _____ perspective.
    A. employment
    B. interactionist
    C. functionalist
    D. conflict

12. The baby boomers have affected _____.
    A. policies of government
    B. how organizations are working
    C. services provided
    D. all of the above

13. The _____ perspective would be interested in looking at the labels we apply to different age groups.
    A. functionalist
    B. conflict
    C. interactionist
    D. all of the above

14. Which of the following is NOT a role being taken on by older adults today?
    A. Volunteerism
    B. New employment
    C. Learning computer skills
    D. None of the above

15. A relatively recent development related to age is _____.
    A. volunteerism of older adults
    B. grandparents raising grandchildren
    C. health care
    D. care-giving

16. _____ is an organization made up of older persons who are retired business professionals who want to help others in their profession be successful.
    A. AARP
    B. Gray Panthers
    C. Senior Corps
    D. Senior Business Partners

17. _____ is an ideology that justifies prejudice or discrimination based on age.
    A. Ageology
    B. Age profiling
    C. Ageism
    D. None of the above

18. Which of the following statements reflects an accurate research finding related to sexual behavior and aging?
    A. Sexual interest and activity tends to decline with age.
    B. Sex remains an important aspect of the lives of aging men and women.
    C. Elderly men are more interested in sex than women of the same age.
    D. All of the above

19. Which of the following is an example of a stereotype of the elderly?
    A. Frail
    B. Incapable
    C. Asexual
    D. All of the above

20. Once a popular view of aging, the _____ theory has now been challenged with the _____ theory.
    A. disengagement; activity
    B. activity; disengagement
    C. decline; sustenance
    D. decline; additive

Essay Question
Answer the following questions using complete sentences in a well-developed essay.

1. List three stereotypes you hear about older people. Explain the basis for each stereotype.

ANSWER KEY

The following provides the answers and references for the practice test questions.

Multiple Choice:

#	Ans	LO	Reference
1.	B	LO 1	textbook, p. 355
2.	D	LO 1	textbook, pp. 355–356
3.	A	LO 1	textbook, pp. 356–358
4.	C	LO 1	textbook, p. 355; video segment 2
5.	A	LO 1	textbook, pp. 355–356; video segment 2
6.	B	LO 1	textbook, p. 361; video segment 2
7.	C	LO 1	textbook, pp. 360–361; video segment 2
8.	D	LO 1	video segment 2
9.	D	LO 1	video segment 2
10.	A	LO 2	video segment 3
11.	C	LO 2	video segment 3
12.	D	LO 2	video segment 3
13.	C	LO 2	video segment 3
14.	D	LO 3	video segment 5
15.	B	LO 3	video segment 5
16.	C	LO 3	video segment 5
17.	C	LO 4	textbook, p. 371
18.	D	LO 4	textbook, p. 370
19.	D	LO 4	textbook, p. 371
20.	A	LO 4	textbook, pp. 372–373

Essay Question

Answers should include the following types of statements/points:

1. ....LO 3 ....textbook, p. 371
    Answer Explanation:
    - Nonsexual
    - Don't like to be around children, younger people
    - Want to only associate with same-age peers
    - Loss of memory
    - Cannot work
    - Society tends to portray older people as inactive and less competent in the workforce.
    - A mandatory retirement age is the result of such stereotypical thinking.

LESSON CONTRIBUTORS

Nana Apt, Professor of Sociology, University of Ghana, Legon, Ghana

Diane R. Brown, Professor of Sociology, Wayne State University, Detroit, MI

Susan Brown Eve, Professor of Sociology, University of North Texas, Denton, TX

Hiram J. Friedsam, Professor Emeritus, University of North Texas, Denton, TX

Arlie Russell Hochschild, Professor of Sociology, University of
California—Berkeley, Berkeley, CA

Ted Pohrte, Educational Consultant, Dallas County Community College District,
Dallas, TX

Nathaniel Eugene Terrell, Chair, Sociology and Anthropology, Emporia State
University, Emporia, KS

Bette Woody, Professor of Sociology, University of Massachusetts, Boston, MA

Dale E. Yeatts, Chair, Department of Sociology, University of North Texas,
Denton, TX

Special acknowledgement and thanks go to:
Garland High School, Garland, TX
Mesquite Social Service, Mesquite, TX
Senior Fest—Eastfield College, Mesquite, TX
AARP, Washington, D.C.

Lesson 15

Deviance and Social Control

OVERVIEW

Sociologist Howard S. Becker identified deviance this way: It is not the act itself, but the reactions to the act, that make something deviant. In exploring deviance, it is important to examine the cultural context in which an act occurs. What may be deviant behavior in one culture or group may be normal behavior in another culture or group.

Sociologists use the term *deviance* nonjudgmentally. An example of *deviance* is when people violate norms from time to time like driving over or under the speed limit.

When sociologists use the sociological perspectives to study criminal deviance, they gain a clearer picture of the complexities and issues. While no single sociological perspective can yield a comprehensive explanation of criminal deviance, in combination the perspectives help view criminal deviance more objectively.

Our society uses formal and informal social controls to discourage, prevent, and punish deviance. Informal social controls regulate less threatening forms of behavior in the everyday interactions of people, as when parents attempt to prevent their children from dressing a certain way or piercing body parts. Often a disapproving look or gesture clearly lets us know when our social behavior is considered deviant and is being discouraged.

Formal social controls include police, prisons, and mental hospitals. These are some of society's means of enforcing social norms by applying social control and maintaining order. How effective are our correctional systems? What is it we want to accomplish in those facilities?

In the United States, society tries to control criminal deviance by focusing on incarceration, but the effectiveness of lengthy prison sentences is frequently debated. In this lesson, a warden and a chaplain along with inmates share their opinions about correctional facilities and the effectiveness of society's social control of deviance. As you listen to their comments, think about how our correctional facilities could be more effective? What needs to be accomplished to reduce criminal deviance?

LESSON ASSIGNMENTS

Text: William Kornblum: *Sociology in a Changing World*, Chapter 7, "Deviance and Social Control," pp. 154–180

Video: "Deviance and Social Control" from the series *Exploring Society, Introduction to Sociology*

LESSON GOAL

After completing this lesson, you will be able discuss how society defines and controls deviance.

LESSON LEARNING OBJECTIVES

1. Discuss the sociological aspects of deviance.
2. Discuss criminal deviance using the sociological perspectives.
3. Discuss the role of prisons as a type of formal social control.
4. Describe the dimensions of deviance.

REVIEW

The following steps are intended to help you learn the material in this lesson. To maximize your learning experience:

   a. Scan the Lesson Focus Point questions.
   b. Read the assigned text pages.
   c. View the video.
   d. Write answers to the Lesson Focus Point questions.
   e. Complete the Related Activities assigned by your instructor. If none are assigned, use them to help you review the lesson material.
   f. Take the Practice Test and check your answers with the Answer Key located at the end of the lesson.

LESSON FOCUS POINTS

1. Define *deviance*?
2. What are some examples of deviance in American society?
3. How do sanctions control deviance? Give examples.
4. What is the difference between informal and formal social control? Give examples.
5. Why is the study of deviance central to the science of sociology?
6. What is *social control*? Give examples.
7. How is the dimension of power related to a definition of deviance?
8. How are ascribed statuses a dimension of deviance?
9. What is *stigma*? How is it related to deviance? Give examples.
10. What is the definition of *crime*?
11. How are changing values related to deviance?
12. What are examples of deviant subcultures?
13. What are factors related to imprisonment and social control? Give examples.
14. What are the reasons sociologists give for our prisons?
15. What are *retribution*, *deterrence*, *rehabilitation*, and *incapacitation*?
16. How effective is capital punishment as a deterrent of criminal behavior?
17. What is *plea bargaining*?
18. How effective are prisons?
19. What is *recidivism*?

RELATED ACTIVITIES

1. Write a paper about how degradation ceremonies are used as a means of social control.

2. Have you ever been a participant or inmate in a total institution? If not, think of a book or movie that depicted life in a total institution. From your own experience or from the book or movie, record illustrations of the different aspects of total institutions described in the textbook.

3. What attitudes, practices, norms, or laws exist in your city that do not allow ex-offenders to escape the stigma of their imprisonment? Describe at least three remedies for this situation, that is, three things that could be done to help ex-offenders shed this stigma after they have paid their debt to society.

4. Describe programs in your community that are intended to help inmates transition into society after being released from prison or jail.

5. Write a paper outlining your suggestions to make prisons more effective.

PRACTICE TEST

Multiple Choice
Choose the BEST answer.

1. Deviance is behavior that violates the _____ of society.
   A. rules
   B. laws
   C. norms
   D. all of the above

2. Which of the following is an example of deviance?
   A. Dyeing your hair green
   B. Wearing outrageous clothing
   C. Being adorned with body piercing
   D. All of the above

3. Which of the following is NOT an example of deviance?
   A. Being homeless and living on the street
   B. Being male and wearing multiple pierced ear studs
   C. Mugging someone
   D. Being a high school student

4. The study of deviance is central to sociology because deviance can result in social problems and _____.
   A. bring about social change
   B. exclude certain types of behavior
   C. control our actions
   D. none of the above

5. A disapproving look or scowl is an example of _____.
   A. informal social control
   B. formal social control
   C. legally sanctioned social control
   D. none of the above

6. When the laws of a society are broken, society turns to _____ social control.
   A. legal
   B. formal
   C. informal
   D. sanctioned

7. The early theories explaining deviance were based on _____.
   A. theories of socialization
   B. theories of criminality
   C. biological beliefs
   D. sociology

8. Crime and deviance are the result of social structures failing to operate properly, according to the _____ perspective.
   A. functionalist
   B. conflict
   C. interactionist
   D. biological

9. The various labels that are used to isolate particular people as outsiders can be explored using the _____ perspective.
   A. functionalist
   B. conflict
   C. interactionist
   D. biological

10. Which of the following is NOT part of society's government social control of crime?
    A. Society's legal codes
    B. Operation of judicial, police, and penal institutions
    C. Operation of rehabilitative institutions
    D. Parental rules

11. _____ is the probability that a person who has served a jail term will commit additional crimes and be jailed again.
    A. Recidivism
    B. Retribution
    C. Return rate
    D. Relapse rate

12. The shorter sentences that result from _____ diminish somewhat the size of prison population.
   A. recidivism
   B. retribution
   C. plea bargaining
   D. plea swapping

13. In the video lesson, Warden John Rupert states that prisons are most effective at _____.
   A. deterrence
   B. retribution
   C. rehabilitation
   D. incapacitation

14. In the video lesson, sociologist Claire Renzetti believes that prisons often become _____ for criminal behavior.
   A. hotbeds
   B. learning centers
   C. ceasing mechanisms
   D. affective deterrents

15. In the video lesson, Chaplain Craig McAllister, believes that being in the *big house* often provides younger inmates with more _____.
   A. status
   B. deterrence
   C. retribution
   D. rehabilitation

16. In the video lesson, inmate Artie Santana, feels that rehabilitation that takes place in prison is _____.
   A. entirely up to each inmates initiative
   B. ineffective at best
   C. outdated skill training
   D. learning new crimes

17. With the *three strikes and you're out* law, the longer, mandatory prison sentence would be given _____.
   A. after conviction on the third felony—regardless of the felony
   B. only if the third felony is committed toward law enforcement officers
   C. only if the third felony is a more serious crime than either of the other two conviction
   D. the death penalty on the third conviction for a felony

18. One result of plea bargaining is _____.
   A. reduction of costs in the criminal justice system
   B. increase of costs in the criminal justice system
   C. increase in crime rates
   D. decrease of crime rates

19. Which of the following is an important dimension of deviance?
   A. Power
   B. Control
   C. Social problems
   D. Prisons

20. Which of the following is NOT a dimension of deviance?
   A. Power
   B. Ascribed statuses
   C. Stigma
   D. None of the above

Essay Question
Answer the following question using complete sentences in a well-developed essay.

1. Describe a behavior that is considered deviant in one community but acceptable in another. What social institutions or groups dominate the culture of the first community that account for the behavior being defined as deviant?

ANSWER KEY

The following provides the answers and references for the practice test questions.

Multiple Choice:

1. C.............LO 1 ....textbook, p. 155
2. D.............LO 1 ....textbook, p. 155
3. D.............LO 1 ....textbook, p. 155
4. A.............LO 1 ....textbook, p. 156
5. A.............LO 1 .... video segment 2
6. B.............LO 1 .... video segment 2
7. C.............LO 2 ....textbook, p. 162 and p. 164; video segment 4
8. A.............LO 2 ....textbook, pp. 164–165
9. C.............LO 2 ....textbook, p. 166–167; video segment 4
10. D.............LO 3 ....textbook, p. 169
11. A.............LO 3 ....textbook, pp. 171–173
12. C.............LO 3 ....textbook, p. 175
13. D.............LO 3 .... video segment 5
14. B.............LO 3 .... video segment 5
15. A.............LO 3 .... video segment 5
16. A.............LO 3 .... video segment 5
17. A.............LO 3 .... video segment 5
18. A.............LO 3 ....textbook, p. 175; video segment 5
19. A.............LO 4 ....textbook, p. 157
20. D.............LO 4 ....textbook, p. 157

Essay Question
Answers should include the following types of statements/points:

1. ....LO 1 ....textbook, p. 155–157; video segment 2
   Members of a gang interpreting behavior (car jacking, fighting) as positive or
   "normal" within their "gang" community. Outside of that community, the
   behavior is judged to be "deviant."

   Note: Students' responses need to demonstrate an understanding of deviance
   and why communities define deviance in different ways.

LESSON CONTRIBUTORS

Troy Duster, Professor of Sociology, New York University, New York, NY

Claire Renzetti, Professor and Chair of Sociology, St. Joseph's University, Philadelphia, PA

Nathaniel Eugene Terrell, Chair, Sociology and Anthropology, Emporia State University, Emporia, KS

Henry Tischler, Professor of Sociology, Framingham State University, Sudbury, MA

Special acknowledgement and thanks go to:

    State of Texas, Department of Criminal Justice

    Randy Cuthrell, Inmate, Hutchins Unit, Texas Department of Criminal Justice

    Stephen Lewsader, Inmate, Hutchins Unit, Texas Department of Criminal Justice

    Craig W. McAlister, Chaplain, Hutchins Unit, Texas Department of Criminal Justice

    John Rupert, Warden, Hutchins Unit, Texas Department of Criminal Justice

    Audie Santana, Inmate, Hutchins Unit, Texas Department of Criminal Justice

Lesson 16

Social Institutions: Religion, Family, and Economics

OVERVIEW

Societies differ, but the people who live within them share basic needs—from food and shelter, to explanations about the meaning of life. We turn to social institutions to meet these needs and fulfill the functional requirements for our survival. All social institutions share common characteristics—groups, roles, values, norms, and the fulfillment of needs.

These characteristics can be found in every social institution from economics to religion. Members of society have concerns about life after death and the meaning of suffering and loss. It is through religion that these concerns are addressed.

In this video lesson, you will learn that groups found within religion include congregations, synagogues, denominations, and charitable organizations. Some of the roles within religion include those of rabbi, priest, minister, worshipper, and missionary. Religion teaches specific values that encourage us to live our lives according to the tenets of our faith. Norms of religion provide guidance and set the parameters of our behavior. Religion, like all social institutions, is not static, but adapts in order to address the challenges presented by social changes.

Within any *one* institution, there may be variations in how these common characteristics manifest themselves. In this video lesson, a good example of the variation in the manifestation of characteristics is seen in the social institution of family. In America, there exists a rich diversity of family structures and living arrangements. Traditionally, the nuclear family, consisting of two parents and their children, was considered to be the typical American family structure. Today, many sociologists think of the nuclear family as more inclusive. The diversity of family structures and living arrangements is a testament to the stability in the midst of variation of the family as a social institution.

Also in this video lesson, the social institution of economics is explored as the backbone of every modern society. Without the provision of services and the production, distribution, and consumption of goods, industrialized societies would find it difficult to survive. Like all social institutions, economics must respond to social change. The economy of the United States became much more global following World War II. The manner in which economics adapts to globalization and new technology will affect the availability of products and provide services that are essential to everyday life.

LESSON ASSIGNMENTS

Text: William Kornblum: *Sociology in a Changing World*, Chapter 15, "The Family," pp. 378–407, Chapter 16, "Religion," pp. 409–438, and Chapter 18, "Economic Institutions," pp. 469–497

Video: "Social Institutions: Religion, Family and Economics" from the series *Exploring Society, Introduction to Sociology*

LESSON GOAL

After completing this lesson, you will be able to discuss the functions of the social institutions of religion, family, and economics in terms of meeting society's needs.

LESSON LEARNING OBJECTIVES

1. Using religion as an example, discuss the common characteristics of social institutions.
2. Using family as an example, discuss how the common characteristics of social institutions can vary in how they manifest themselves within the institution.
3. Using economics as an example, discuss how a social institution might respond to various challenges generated by social change.

REVIEW

The following steps are intended to help you learn the material in this lesson. To maximize your learning experience:

   a. Scan the Lesson Focus Point questions.
   b. Read the assigned text pages.
   c. View the video.
   d. Write answers to the Lesson Focus Point questions.
   e. Complete the Related Activities assigned by your instructor. If none are assigned, use them to help you review the lesson material.
   f. Take the Practice Test and check your answers with the Answer Key located at the end of the lesson.

LESSON FOCUS POINTS

1. Describe social institutions in terms of needs, groups and organizations, values, roles, and norms.
2. What are the five requisites that each society must fulfill if it is to survive?
3. Describe unique family structures and how they meet the needs of the family.
4. What are the challenges and trends for family as a social institution?
5. How do the sociological perspectives provide insights about the family as a social institution?
6. What are the definitions of *sacred* and *profane*?
7. How are the characteristics of a social institution exemplified in religion?
8. Explain how religion can be both a source of division and healing?
9. What are the major world religions? Forms of religion?
10. What do these terms have in common? S*imple supernaturalism*, *animism*, *atheism*, *polytheistic*, and *monotheistic*. Give examples for each.
11. How does religion affect social change?
12. What are different examples of religious organizations, sects, denominations, secularization, cults, and religiosity?
13. What are some trends in religion in the United States?
14. What are the advantages of religious pluralism?
15. How does economics function to meet society's needs?
16. What do the terms: *markets*, *subsistence economy*, *multinational corporation*, and *economic globalization* have in common?
17. What changes have facilitated globalization?
18. Is the United States a postindustrial society?
19. What are some of the unresolved social problems related to economics?
20. What are possible sources of worker alienation?
21. What are *corporate actors*? What is their significance?
22. How do sociologists define the term *profession*?

RELATED ACTIVITIES

1. Many religions have a written record of their "set of coherent answers." Pick one of those religions and describe where these answers are found. Explain how they solve some of the "dilemmas of human existence" for their believers. (You need not pick a religion with whose answers you happen to agree.)

---

2. Describe at least two instances in which religion has either generated or thwarted social change. Explain what the change is or was and how the religion affected it and why.

3. If your family is a "blended" family involving step-parents, step-children, or "half" brothers and/or sisters, write about the most difficult situations/problems you have encountered in your "new" family. Your perspective may be that as a parent or child—use whichever role you fulfill. How effectively did you deal with each situation? What has been the most difficult to resolve? What lessons have you learned from being in a "blended" family?

4. Identify a multinational corporation and briefly describe the nature of its business. In which countries does this corporation employ personnel and own capital equipment or buildings? In which countries does it sell its product or service?

5. Interview an older man or an older woman (i.e. your grandparents, great aunts or uncles, etc.) about the first "real" job they had after finishing school. Describe the work, salary, and any requirements that the job demanded before being hired. Then compare that salary to a similar position today.

PRACTICE TEST

Multiple Choice
Choose the BEST answer.

1. Which of the following is NOT a characteristic of social institutions?
   A. Values
   B. Philosophy
   C. Roles
   D. Groups or organizations

2. Which of the following is NOT an example of a group or organization that serves the social institution of religion?
   A. Congregations
   B. Pastors
   C. Synagogues
   D. Denominations

3. Which of the following reflects religion conferring legitimacy on society's norms and values?
   A. Wedding ceremony
   B. Baptism
   C. Bar mitzvahs
   D. All of the above

4. Which of the following is a role often found in the social institution of religion?
   A. Those who prepare communion
   B. Those who make repairs in the building
   C. Those who manage the finances
   D. All of the above

5. Social justice is a _____ within the social institution of religion.
   A. need
   B. value
   C. role
   D. status

6. Which of the following would NOT be a norm often found in religion?
   A. Pro-family rules
   B. Treating people with respect
   C. Singing in the choir
   D. Kindness toward other people

7. Religion often creates _____ where we feel comfortable and where life seems livable.
   A. moral boundaries
   B. moral communities
   C. territorial communities
   D. membership communities

8. A family is a group of people related by _____.
   A. consanguineous attachments
   B. conjugal relations
   C. legal adoption
   D. all of the above

9. Traditionally, the _____ family consists of parents and their children.
   A. ideal
   B. concentric
   C. nuclear
   D. real

10. Today, most sociologists would think of all of the following as examples of nuclear families EXCEPT _____.
    A. mother, father, children
    B. mother, children
    C. father, children
    D. single mother

11. Which of the following is NOT a reason for the increasing numbers of single-parent families?
    A. Divorce
    B. Economic pressures
    C. No longer a stigma attached
    D. Death of spouse

12. The nuclear family in which a person is born and socialized is termed the family of _____.
    A. orientation
    B. procreation
    C. origin
    D. kinship

13. Which of the following statements reflects a demographic trend revealed by sociologists?
    A. Traditional households (mother, father, and children) are declining in number.
    B. There has been an increase in proportion of married couples without children.
    C. There has been an increase in the single-parent family.
    D. All of the above.

14. Which of the following is NOT a reason discussed in the video for the increase in intergenerational families?
    A. Divorced children returning home
    B. Single parents moving home
    C. Older parents needing assistance
    D. Desire for extended family

15. Which of the following is NOT a change that has greatly facilitated globalization in recent years?
    A. Innovations and advances in transportation
    B. Regional and state tariffs
    C. Global economic liberalization
    D. Use of market incentives

16. Which political economics ideology is represented in the United States?
    A. Welfare capitalism
    B. Democratic socialism
    C. Mercantilism
    D. *Laissez-faire* capitalism

17. Which of the following trends is related to the increases in the demands for part-time workers?
    A. Contracting with freelance workers
    B. Employers not paying health care benefits
    C. Preventing employees from joining labor unions
    D. All of the above

18. Which subfield of sociology would be concerned with the social organization of work and the types of interactions that occur in the workplace?
    A. Organizational sociology
    B. Industrial sociology
    C. Employment sociology
    D. Workplace sociology

19. Which of the following is a reason economics is so critical to the rest of society?
    A. It is the system through which we provide food.
    B. It is the system through which we provide shelter.
    C. It is the system through which we establish global market links.
    D. All of the above

20. After World War II, the economy of the United States became much more
    _____ in nature.
    A. limited
    B. critical
    C. contributing
    D. global

Essay Questions
Answer the following questions using complete sentences in a well-developed essay.

1. Describe at least two rituals of a particular religion and the shared meanings connected with those rituals.

2. Describe the forms of religion.

3. How was the labor divided in your family of orientation? Explain how the pattern set there affects your family roles today or the roles you are likely to play in the future.

4. Describe the changes that have greatly influenced globalization.

ANSWER KEY

The following provides the answers and references for the practice test questions.

Multiple Choice:

1. B..............LO 1..textbook, p. 410; video segment 2
2. B..............LO 1..textbook pp. 410–411; video segment 2
3. D..............LO 1..textbook, pp. 410–411
4. D..............LO 1..video segment 2
5. D..............LO 1..video segment 2
6. D..............LO 1..video segment 2
7. B..............LO 1..video segment 2
8. D..............LO 2..textbook, pp. 381–382
9. C..............LO 2..textbook, p. 382; video segment 3
10. D..............LO 2..textbook, pp. 381–383; video segment 3
11. D..............LO 2..video segment 3
12. A..............LO 2..textbook, p. 382
13. D..............LO 2..textbook pp. 382–383
14. D..............LO 2..video segment 3
15. B..............LO 3..textbook, p. 474
16. A..............LO 3.. textbook, p. 478 and pp. 480–481
17. D..............LO 3..textbook, p. 484
18. B..............LO 3..textbook, p. 488
19. D..............LO 3..textbook, pp. 470–471; video segment 4
20. D..............LO 3..video segment 4

Essay Questions

Answers should include the following types of statements/points:

1. ....LO 1..textbook, pp. 410; video segment 2
    - Weddings
    - Baptisms
    - Bar mitzvahs
    - Confirmations
    - Torah (Jews)
    - Bible (Christians)
    - Koran (Muslims)
    - Holy communion

2.   ....LO 1 ....textbook, pp. 414–419

Form of Religion:	Description:
Simple Supernaturalism	A form of religion in which there is no discontinuity between the world of the senses and the supernatural; all natural phenomena are part of a single force
Animism	A form of religion in which all forms of life and all aspects of the earth are inhabited by gods or supernatural powers
Theism	A form of religion in which gods are conceived of as separate from humans and from other living things on the earth, although the gods are in some way responsible for the creation of humans for their fate
Polytheism	A form of theism in which there are numerous gods, all of whom occupy themselves with some aspect of the universe and of human life
Monotheism	A form of theism that is centered on belief in a single all-powerful God who determines human fate and can be addressed through prayer
Abstract Ideals	A form of religion that is centered on an abstract ideal of spirituality and human behavior
Civil Religion	A collection of beliefs, and rituals for communicating those beliefs, that exists outside religious institutions

3.   ....LO 2 ....textbook, pp. 379–382

Father:	Mother:
• Take out garbage	• Cooking
• Mow yard	• Cleaning
• Barbecue	• Washing
• Punishment reinforcement	• Ironing
• Provide income	• Sewing
	• Child rearing

4. ....LO 3..textbook, pp. 471–476
These aspects of globalization have been greatly facilitated in recent decades by three important changes:

- Innovations and advances in transportation (e.g. jet travel and air freight) and communications (computers and the Internet), which allow extremely rapid exchanges of people, funds, and capital between regional markets.
- Global economic liberalization—especially the reduction of national tariffs that may create barriers to the free flow of trade and investment funds—are encouraged through global trade institutions, particularly the General Agreement on Tariffs and Trade (GATT) and the World Trade Organization (WTO) in the case of world trade in goods and services, and the International Monetary Fund (IMF) in the case of global finance and capital flows.
- Use of market incentives that encourage people to rely on supply and demand rather than government-enforced quotas, regulations, or treaties to regulate economic behavior—to speed up the development of production and trade everywhere in the world (Rajan, 2001)

LESSON CONTRIBUTORS

Bruce A. Chadwick, Professor of Sociology, Brigham Young University, Provo, UT

Carol Deming Chenault, Instructor of Sociology, Calhoun Community College, Decatur, Alabama

Gordon Fellman, Professor of Sociology, Brandeis University, Waltham, MA

Phillip B. Gonzales, Associate Professor of Sociology, University of New Mexico, Albuquerque, NM

Arlie Russell Hochschild, Professor of Sociology, University of California—Berkeley, Berkeley, CA

Paul Magee, Professor of Sociology, North Lake College, Irving, TX

Rudy Ray Seward, Professor and Associate Chair of Sociology, University of North Texas, Denton, TX

Timothy L. Sullivan, Professor of Sociology and Anthropology, Cedar Valley College, Lancaster, TX

Brian Uzzi, Professor of Sociology, Northwestern University, Evanston, IL

Bette Woody, Professor of Sociology, University of Massachusetts, Boston, MA

Special acknowledgement and thanks go to:
The Kaplan Family, Los Angeles, CA
The White Family, Flower Mound, TX
Janna Oehlschlaeger, Dallas, TX
LaDena Thomas, Duncanville, TX
Audrey Bulls and Kathleen Young, Dallas, TX
Body of Christ Family Church, Red Oak, TX
DFW Hindu Temple Society, Irving, TX
Karma Thegsum Choling, Tibetan Buddhist Meditation Center, Dallas, TX
North Texas Mosque, Richardson, TX
Santa Clara Catholic Church, Dallas, TX
Temple Emanu-El Synagogue, Dallas, TX

Lesson 17

Social Institutions: Politics and Education

OVERVIEW

Every day, we interact with social institutions. These interactions take place as we earn our living, nurture our children, and further our education. Social institutions influence our actions and our relationships with others. They provide the structure within which we live our lives.

In this video lesson, you'll learn how Americans interact with the social institution of education throughout the school year. Our children attend public school for seven hours a day, five days a week. School zones affect morning and afternoon traffic patterns. School schedules influence how we structure our personal time and activities—from socializing with friends and family members to planning vacations.

The social institution of education keeps society functioning by passing on essential knowledge, skills, and information from one generation to the next. Education also functions to meet our need for lifelong learning, which becomes more critical with rapidly changing technology. Included in lifelong learning is the need for individuals to be retrained so they can remain in the workforce longer.

Like other social institutions, American education faces challenges. Some of these challenges include discipline, culturally biased tests, social promotion, and grade inflation.

One of society's greatest needs is to establish a social structure of power and authority to counteract disorder and chaos. This structure is called politics, one of many social institutions within American society. The social institution of politics meets society's needs for order, structure, purpose, and faces many challenges as all social institutions do.

In this video lesson, you'll hear about one of the most unique challenges in American politics from the year 2000. This particular challenge didn't come from a foreign power, but rather from issues within the institution. At stake was the integrity of the electoral process to determine who would become President of the United States.

Also in this video, you will meet Anna. She is a twenty-one-year-old college student who works part-time in an office. She will help you understand how your life is influenced by a number of social institutions on a daily basis.

By becoming aware of how frequently we interact with social institutions, we can begin to appreciate the critical role they play in our daily lives.

LESSON ASSIGNMENTS

Text: William Kornblum: *Sociology in a Changing World*, Chapter 17, "Education," pp. 440–467, and Chapter 19, "Politics and Political Institutions," pp. 499–525

Video: "Social Institutions: Politics and Education" from the series *Exploring Society, Introduction to Sociology*

LESSON GOAL

After completing this lesson, you will be able to discuss the functions of the social institutions of politics and education in terms of meeting society's needs.

LESSON LEARNING OBJECTIVES

1. Using education as an example, discuss the common characteristics of a social institution.
2. Using politics as an example, discuss how a social institution might respond to internal issues.
3. Explain how social institutions influence daily choices and actions.

REVIEW

The following steps are intended to help you learn the material in this lesson. To maximize your learning experience:

   a. Scan the Lesson Focus Point questions.
   b. Read the assigned text pages.
   c. View the video.
   d. Write answers to the Lesson Focus Point questions.
   e. Complete the Related Activities assigned by your instructor. If none are assigned, use them to help you review the lesson material.
   f. Take the Practice Test and check your answers with the Answer Key located at the end of the lesson.

LESSON FOCUS POINTS

1. How do the common characteristics of social institutions apply to education?
2. Provide examples of *hidden curriculum*.
3. Why is *lifelong learning* important?
4. What are some of the *roles, functions, groups*, and *values* found in education?
5. What are some suggested ways we can combat grade inflation and degree inflation?
6. Using the sociological perspectives, explain the social institution of education.
7. According to the textbook, what are the primary findings of James Coleman in his study, *The Adolescent Society*?
8. What is the difference between *educational attainment* and *educational achievement*?
9. What happens when a child is a product of educational tracking?
10. What is a commonly cited reason for dropping out of school?
11. What is the *reproductive theory* as it relates to the social institution of education?
12. In education, what is meant by *total immersion*?
13. According to the textbook, what are the findings from research conducted by Sandra Baum?
14. How does the *human capital theory* relate to education?
15. Why are schools considered to be bureaucracies?
16. What are some issues related to comprehensive school reform?
17. How is class size related to educational achievement?
18. What is *school choice*? The *home school movement*?
19. According to the textbook, what did Jonathan Kozol's research reveal about racial and class segregation?
20. What is the function of the social institution of politics?
21. What challenges surfaced in the presidential election of 2000?
22. What are *power* and *authority,* as they relate to politics?
23. What are the differences between a political institution and the social institution of politics?
24. What constitutes "the state" in separation of church and state?
25. What is *traditional authority* versus *legal authority* versus *charismatic authority*?
26. What is society's political culture?
27. How is *nationalism* established?
28. What are the characteristics of modern political systems?
29. What conditions must be met in a democratic political system?

30. What are examples of *typologies of regimen*?
31. What are the differences between the power and pluralist models in politics?
32. Why is lobbying important?
33. How has media affected politics?
34. Why are *opinion leaders* important? Give examples.
35. According to the lesson video, why were many voters in Palm Beach County disqualified and kept from voting?
36. According to the lesson video, how does Anna's life interact with social institutions during the course of a day?

RELATED ACTIVITIES

1. Describe two typical situations in your family, or in a close relationship, in which politics operate.

2. What is your opinion about the effectiveness of the political institutions of your society in dealing with problems of inequality and injustice? Cite at least three pieces of evidence to support your view.

3. Would you describe the power structure in your city or town as being a power elite or a pluralist structure? Give arguments to support your view.

4. List at least three interest groups that are involved in lobbying. Describe the goal or purpose of each group, and tell how many members are in each.

5. Thinking back on your high school days, were you or your friends part of an "adolescent society," as described by James Coleman and other sociologists? Did you experience a "compulsive conformity" and loyalty to a peer group? If you cannot identify with these questions for yourself, do you see these phenomena happening with your children or other teenagers?

6. List and explain three positive and three negative factors you experienced in your education (school).

7. Do you agree with the author of the textbook that the expectations that education can solve social problems seem impossible to meet? Explain why or why not.

PRACTICE TEST

Multiple Choice
Choose the BEST answer.

1. Which of the following is NOT a function of the social institution of education?
   A. Transmission of knowledge
   B. Preparation of young people for adulthood
   C. Preparation of adults for new roles
   D. None of the above

2. Which of the following is true about the goals of education for citizenship within American society?
   A. Schools have been expected to teach values but not to shape society itself.
   B. Schools have been expected to create equal opportunity and prepare new generations of citizens to function in society.
   C. That education should help children maintain their cultural identity is almost universally accepted.
   D. Nearly all school districts throughout the country have adopted the goal of encouraging critical thinking skills in their students.

3. The informal, subtle social norms that we learn from education are sometimes referred to as the _____.
   A. primary function of education
   B. hidden curriculum
   C. hidden agenda
   D. unwritten curriculum

4. The term used for the number of years of school completed is educational _____.
   A. attainment
   B. achievement
   C. competency
   D. background

5. Which of the following is NOT a problem with regard to educational attainment?
   A. Tracking systems
   B. Social movements
   C. Degree inflation
   D. Dropping out

6. Which of the following is NOT a value promoted in the social institution of education?
   A. Responsibility
   B. Respect for others
   C. Sharing
   D. Playing football

7. Which of the following would be influenced by an interaction with the social institution of education?
   A. Morning and afternoon traffic patterns
   B. How we structure personal time and activities
   C. When we take the family on vacation
   D. All of the above

8. Which of the following would be an example of lifelong learning?
   A. Retraining when a person loses a job
   B. Learning how to use a computer
   C. Learning new skills to cope with changes in one's life
   D. All of the above

9. A Parent-Teacher Association is an example of a _____ found within the social institution of education.
   A. role
   B. norm
   C. group
   D. need

10. It is through _____ that society establishes a social structure of power and authority.
    A. education
    B. politics
    C. economics
    D. law

11. _____ are specialized organizations that attempt to influence elected and appointed officials on specific issues.
    A. Interest groups
    B. Political groups
    C. Political sectors
    D. Interest sectors

12. Which of the following is an example of a lobby group?
    A. Trade unions
    B. AARP
    C. NAACP
    D. All of the above

13. Today, _____ has become the dominant media of political communication.
    A. radio
    B. Internet
    C. television
    D. both A and C

14. Television's dominance and influence on political campaigns have led to the movement for _____.
    A. free "air" time for candidates
    B. campaign finance reform
    C. campaign contributions
    D. a more informed electorate

15. According to the video lesson, which of the following is a question generated as a result of the 2000 presidential election?
    A. What happens when technology fails?
    B. What happens if there is a tie in the votes?
    C. What happens if groups of people feel disenfranchised by the process?
    D. All of the above

16. The right to vote is a _____ in the social institution of politics.
    A. need
    B. value
    C. norm
    D. role

17. According to the video lesson, which of the following daily activities would constitute an interaction with the social institution of economics?
    A. Shutting off your alarm clock
    B. Toasting bread in a toaster
    C. Frying an egg
    D. All of the above

18. We most frequently interact with the social institution of economics through our _____.
    A. jobs
    B. education
    C. voting
    D. attending church

19. In the video lesson, Anna interacts with the social institution of education when she attends a _____.
    A. technical training session
    B. community college
    C. university
    D. on the job training session

20. Attending a city council meeting in our community is an example of how we might interact with the social institution of _____.
    A. politics
    B. economics
    C. education
    D. law

Essay Questions
Answer the following questions using complete sentences in a well-developed essay.

1. List and explain three positive and three negative factors you experienced in your education (school).

2. Describe two typical situations in your family, or in a close relationship, in which politics operates.

ANSWER KEY

The following provides the answers and references for the practice test questions.

Multiple Choice:

1. D.............LO 1...textbook, p. 442; video segment 2
2. B.............LO 1...textbook, pp. 446–450; video segment 2
3. B.............LO 1...textbook, p. 443
4. A.............LO 1...textbook, p. 450
5. B.............LO 1...textbook, p. 450–453
6. D.............LO 1...video segment 2
7. A.............LO 1...video segment 2
8. D.............LO 1...video segment 2
9. D.............LO 1...video segment 2
10. B.............LO 2...textbook, pp. 500–501; video segment 3
11. A.............LO 2...textbook, p. 516
12. D.............LO 2...textbook, p. 516
13. C.............LO 2...textbook, pp. 517–519
14. B.............LO 2...textbook, pp. 517–518
15. D.............LO 2...video segment 3
16. B.............LO 2...video segment 3
17. D.............LO 3...video segment 4
18. A.............LO 3...video segment 4
19. B.............LO 3...video segment 4
20. A.............LO 3...video segment 4

Essay Questions
Answers should include the following types of statements/points:

1. ....LO 1...textbook, pp. 441–466
    Answer Explanation:

Positive:	Negative:
• Parental involvement	• Bureaucratic with hierarchy
• Small classes	• Little or no control
• Personal relationships among students, teachers, and administrators	• Large classes
• Everyone knew name	• Can be isolated
• "Open" classroom	• Teachers impersonal
• Caring teacher(s)	• Some activities (band, music) not supported
• Extracurricular activities	• Teachers' "favorites"
• Good grades	

2. ....LO 2, 3 ....textbook, pp. 499–524; video segment 4
Answer Explanation:
- Majority "rules"
- Power/authority rests with parents
- Rules set by parents
- Order of sitting arrangement at dinner

LESSON CONTRIBUTORS

Bruce A. Chadwick, Professor of Sociology, Brigham Young University, Provo, UT
Susan Brown Eve, Professor of Sociology, University of North Texas, Denton, TX
Phillip B. Gonzales, Associate Professor of Sociology, University of New Mexico,
    Albuquerque, NM
Arlie Russell Hochschild, Professor of Sociology, University of
    California—Berkeley, Berkeley, CA
Philip Luck, Assistant Professor of Sociology, Lyndon State College, Lyndonville, VT
Theresa A. Martinez, Associate Professor of Sociology, University of Utah, Salt
    Lake City, UT
Brian Uzzi, Professor of Sociology, Northwestern University, Evanston, IL
Bette Woody, Professor of Sociology, University of Massachusetts, Boston, MA

Special acknowledgement and thanks go to:
    Mary Balser, Desoto, TX

Lesson 18

Health and Medicine

OVERVIEW

Health and medicine mean many different things to different people. Some take it for granted that it will ensure them a long, healthy life. To others, it is a luxury they cannot afford. Sociologists know that this important social institution involves more than biology—it also involves cultural beliefs, lifestyles, and social class.

Some health care professionals may view the practice of medicine as their calling in life but patients often think important medical decisions are determined by impersonal economic policies and government regulations. Sociologists explore these multidimensional aspects of health and medicine by using the sociological perspectives. In doing so, they provide valuable insights about the complexities of this social institution.

While America is a tapestry of many cultures and practices, American health care continues to emphasize *curative health,* which focuses on the diagnosis and treatment of illness, rather than *preventive health,* which stresses the need to live our lives to optimize health, through proper diet, exercise, and annual examinations. In this video lesson you'll hear about some exotic—even mysterious—alternative healing approaches, which enrich the offerings of American healthcare.

Age, gender, ethnicity, and geographical location have a part in determining our health and the level of care available to us. These are the demographic factors studied by epidemiologist Kimberly Peters, a sociologist working for the New Mexico Department of Health. In this video lesson, you'll hear Kimberly talking about the "Women's Health Profile, 2001," a study which will hopefully lead to the improvement of women's health care in New Mexico.

Over the past fifty years, advances in medical technology have greatly improved the diagnostic process and made many surgeries less invasive. Advances have also been made in pharmaceutical drugs—offering cures for many diseases and relief from the symptoms of many other disorders.

New technologies and drugs have contributed to the soaring cost of health care and created a two-tiered system of medical care—superior care for those who can afford the cost, and inferior care for those who can't. This inequity often creates moral and ethical issues.

With greater emphasis on cost containment and insurance plans setting limits on services and fees, the question remains: is the social institution of health and medicine doing a better or worse job of serving patients today? And what about the future?

LESSON ASSIGNMENTS

Text: William Kornblum: *Sociology in a Changing World*, Chapter 20, "Technology, Environment, and Medicine," pp. 540–548

Video: "Health and Medicine" from the series *Exploring Society, Introduction to Sociology*

LESSON GOAL

After completing this lesson, you will comprehend the sociological significance of medicine and health care.

LESSON LEARNING OBJECTIVES

1. Use the sociological perspectives to examine the social institution of health and medicine.
2. Discuss how age, gender, race/ethnicity, and socioeconomic status are related to health and illness.
3. Describe the sociological implications of the future of the U.S. health care system.
4. Discuss the sociological significance of science.
5. Explain the significance of medical technology.
6. Identify some alternative health care methods.

REVIEW

The following steps are intended to help you learn the material in this lesson. To maximize your learning experience:

a. Scan the Lesson Focus Point questions.
b. Read the assigned text pages.
c. View the video.
d. Write answers to the Lesson Focus Point questions.
e. Complete the Related Activities assigned by your instructor. If none are assigned, use them to help you review the lesson material.
f. Take the Practice Test and check your answers with the Answer Key located at the end of the lesson.

LESSON FOCUS POINTS

1. What are the two approaches most often followed by sociologists who study scientific institutions?
2. What are the norms of science?
3. What improvements emerged during the nineteenth century regarding public health?
4. What are the stages of development of hospitals?
5. How does *hypertrophy* relate to health care?
6. What is the significance of *medical sociology*?
7. What are recent challenges faced by medical sociologists?
8. What are some ethical considerations related to health and medicine?
9. What is the central focus of the functionalist perspective regarding health and medicine?
10. How does the functionalist perspective define the *sick role* and the *physicians role*?
11. What is the central feature of the conflict perspective regarding health and medicine?
12. Which sociological perspective would examine gender differences in treatment?
13. What is the central focus of the interactionist perspective regarding health and medicine?
14. What is the difference between *curative medicine* and *preventive medicine*?

15. Discuss some alternative health care methods practiced in the United States such as: *acupuncture, ayurveda, curandisimo,* and *curanderas.*
16. What is an epidemiologist?
17. What social factors would be studied by an epidemiologist?
18. What was the goal of the research report from New Mexico presented in the video lesson?

RELATED ACTIVITIES

1. What are the arguments for and against the right to die? With which of these views do you agree and why? Write a brief essay presenting both views about this critical health care issue.

2. Discuss the major issues facing insurance companies today.

3. Explain why patients need to become involved in preventive medicine.

4. Explain the role of a social epidemiologist.

PRACTICE TEST

Multiple Choice
Choose the BEST answer.

1. The central focus of the functionalist perspective when examining health and medicine is _____.
   A. social class differences
   B. access to health care
   C. maintaining healthy people
   D. personal experiences

2. According to the functionalist perspective, two central issues in health and medicine are _____ and _____.
   A. the sick role; the physician's role
   B. accessibility; social control
   C. social roles; access
   D. regaining health; access to health care

3. Access to health care is the focal point of the _____ perspective.
   A. functionalist
   B. conflict
   C. interactionist
   D. marginal

4. The _____ perspective examines how health and medicine are socially constructed.
   A. interactionist
   B. functionalist
   C. conflict
   D. feminist

5. The statement, "People can have the same medical condition yet respond in very different ways" is aligned with the _____ perspective.
   A. feminist
   B. functionalist
   C. conflict
   D. interactionist

6. A(n) _____ seeks out the causes of illnesses and diseases and how they are distributed in population groups.
   A. sociologist
   B. ethnographer
   C. epidemiologist
   D. medical sociologist

7. The goal of the research featured in the video was to _____.
   A. raise health standards in New Mexico
   B. upgrade health care in New Mexico
   C. improve health care for women in New Mexico
   D. present a national picture of health care

8. Which of the following social issues might be examined through social epidemiology?
   A. Racism
   B. Gender bias
   C. Poverty
   D. All of the above

9. Which of the following is true of the health care in the United States in the past fifty years?
   A. Medical technology has greatly improved.
   B. Diagnostic processes have made many surgeries less invasive.
   C. Pharmaceutical drug advances provide relief for many diseases.
   D. All of the above

10. For health insurance providers, the most significant issue is _____.
   A. access to health care professionals
   B. certification and education of providers
   C. balancing cost of care with quality of care
   D. numbers of patients seeking care

11. Which of the following is the function of the institutions of science?
   A. Promote public health
   B. Collaboration between universities and private industry
   C. Extend knowledge by means of a specific set of procedures
   D. Invention

12. _____ means that the truth of scientific knowledge must be determined by impersonal criteria of the scientific method.
   A. Universalism
   B. Disinterestedness
   C. Impartiality
   D. Common publication

13. Which of the following is NOT a norm of science?
   A. Universalism
   B. Common ownership/publication
   C. Disinterestedness
   D. Applicationism

14. Which of the following improvements during the nineteenth century contributed positively to public health practices?
   A. Separation of drinking water from waste water
   B. Awareness of importance of hygiene
   C. Knowledge of need for sterilization
   D. All of the above

15. In the late nineteenth century, _____ evolved for the delivery of health care.
    A. community clinics
    B. hospitals
    C. asylums
    D. none of the above

16. Which of the following was NOT a result of the increased awareness of public health?
    A. Decrease in infant mortality
    B. Life expectancy decreased
    C. Births began to outnumber deaths
    D. All of the above

17. Some critics who say that the American medical health care system has expanded to a size and complexity at which it has become dysfunctional. This statement is a reflection of _____.
    A. megalopolis
    B. hyperhealth care
    C. hypertrophy
    D. hyperextension

18. _____ medicine focuses on taking the initiative in maintaining health.
    A. Curative
    B. Preventive
    C. Alternative
    D. Scientific

19. One of the most frequently practiced forms of traditional Chinese medicine in the United States is _____.
    A. curative medicine
    B. preventive medicine
    C. acupuncture
    D. ayurveda

20. _____ is the art of Mexican folk healing.
    A. Acupuncture
    B. Yoga
    C. Ayurveda
    D. Curandisimo

Essay Questions
Answer the following questions using complete sentences in a well-developed essay.

1. Discuss the public health improvements that occurred in the nineteenth century.

2. Discuss the differences between curative and preventive health care giving an example of each.

ANSWER KEY

The following provides the answers and references for the practice test questions.

Multiple Choice:
1. C ....LO 1 .... video segment 2
2. A ....LO 1 .... video segment 2
3. B ....LO 1 .... video segment 2
4. A ....LO 1 .... video segment 2
5. D ....LO 1 .... video segment 2
6. C ....LO 2 .... video segment 4
7. C ....LO 2 .... video segment 4
8. D ....LO 2 ....textbook, p. 544
9. D ....LO 3 .... video segment 6
10. C ....LO 3 .... video segment 6
11. C ....LO 4 ....textbook, pp. 528–529
12. A ....LO 4 ....textbook, pp. 529–530
13. D ....LO 4 ....textbook, pp. 529–531
14. D ....LO 5 ....textbook, pp. 540–542
15. B ....LO 5 ....textbook, pp. 542–543
16. B ....LO 5 ....textbook, p. 542
17. C ....LO 5 ....textbook, pp. 543–544
18. B ....LO 6 .... student course guide lesson overview, pp. 173–174
19. C ....LO 6 .... video segment 3
20. D ....LO 6 .... video segment 3

Essay Questions
Answers should include the following types of statements/points:

1. ....LO 5 ....textbook, pp. 540–543
    *   Information about hygiene and sterilization became known
    *   Awareness of the need to separate drinking water from waste water
    *   Awareness of sanitation techniques

---

2. ....LO 6..student course guide lesson overview, pp. 173–174
- Curative medicine is treating a medical problem after one becomes ill, such as going to a doctor because you have the flu, receiving medication to relieve symptoms of illness.
- Preventive health care is taking steps to avoid illness, such as getting annual physical examination, dental check-ups, living a life that optimizes health (diet, exercise).

LESSON CONTRIBUTORS

Diane R. Brown, Professor of Sociology, Wayne State University, Detroit, MI

Don Cornwell, Acupuncturist, New Mexico School of Natural Therapeutics, Albuquerque, NM

Susan Brown Eve, Professor of Sociology, University of North Texas, Denton, TX

Chitra Giauque, Yoga Instructor, Ayurvedic Institute, Albuquerque, NM

Roger E. Herman, Chief Executive Officer and Strategic Business Futurist, The Herman Group, Greensboro, NC

Kimberley Peters, Statistical Research Director, New Mexico Vital Records and Health Statistics, New Mexico Department of Health, Santa Fe, NM

Jessica Ruehrwein, Program Director, Girls Incorporated of Santa Fe, Santa Fe, NM

Richard Scotch, Professor of Sociology, University of Texas at Dallas, Dallas, TX

Special acknowledgement and thanks go to:

Ron Austin, Social Worker, Baylor Senior Health Center, Dallas, TX

Lisa Garner, MD, Dermatologist, Garland, TX

Jane E. Nokleberg, M.D., Dallas, TX

Susana Quiroga, McKinney, TX

Hinke Schroen, Dallas, TX

David Sigel, LMSW-CCM, Care Coordinator/Social Work Supervisor

Eliseo Torres, Author and Curandero, Albuquerque, NM

Curves for Women, Santa Fe, NM

Lesson 19

Communications Media and Technology

OVERVIEW

The social institution of communications media is a powerful force in America. It informs, instructs, and entertains us. In the case of news organizations, its primary power rests not in telling us what to think, but in making choices about news coverage, which in turn influences how we interpret and perceive events around us.

Words and images shape our ideas and our opinions. While the communications media reveals a world beyond our immediate community, its power and influence may go unnoticed. Who should choose what news to cover? At what level should these decisions be made? And what are the ramifications of the choices?

In this video lesson, you will explore the social role of communications media during war to better understand how public opinion is shaped. Before television, radio, and film, our sources for news about war were newspapers and word-of-mouth. But by the time World War II began, technological advances made it possible for radio correspondents to report from the front lines and for the general public to view newsreel footage of battles. On September 11, 2001, the communications media broadcast material from a variety of sources to reveal the horrifying images that marked the beginning of America's war on terrorism.

A hybrid form of television entertainment has bridged the gap between fact and fiction. So-called *reality television* shows people engaged in activities that range from breaking the law to self-imposed exile from civilization. But how "real" is reality television? Reality television has a strong appeal for a large part of the viewing public. But the fact remains that within these programs, choices are made in how the footage is edited to make the scenes more dramatic. This makes it difficult to perceive what is real—and what is not—in reality television.

Technology has touched every aspect of our lives. Personal computers, the Internet, and cellular phones are part of our culture. While we welcome the advantages new technology brings, we must often accept the negative effects with the positive. In this video lesson, you will explore the paradox of *technological dualism*, in a story about our relationships with cell phones. You will also learn about *environmental stress*, which often is a negative byproduct of our technological society.

LESSON ASSIGNMENTS

Text: William Kornblum: *Sociology in a Changing World*, Chapter 20, "Technology, Environment, and Medicine," pp. 527–540

Video: "Communications Media and Technology" from the series *Exploring Society, Introduction to Sociology*

LESSON GOAL

After completing this lesson, you will be able to explain how communication media and technology contribute to social change.

LESSON LEARNING OBJECTIVES

1. Using war as an example, discuss the social role of the communications media.
2. Explain how groups within the communications media influence public opinion.
3. Discuss the societal impact of technology.

REVIEW

The following steps are intended to help you learn the material in this lesson. To maximize your learning experience:

a. Scan the Lesson Focus Point questions.
b. Read the assigned text pages.
c. View the video.
d. Write answers to the Lesson Focus Point questions.
e. Complete the Related Activities assigned by your instructor. If none are assigned, use them to help you review the lesson material.
f. Take the Practice Test and check your answers with the Answer Key located at the end of the lesson.

LESSON FOCUS POINTS

1. What is *technology*?
2. What are the dimensions of technology?
3. What is *technological dualism*? Give examples.
4. What are examples of catalysts for social change?
5. What is *technology assessment*?
6. What are some advances in technology that have brought forth large industries?
7. What do some sociologists think will cause a new long boom in technology?
8. What is *pollution*?
9. What is *environmental stress*?
10. What new improvements emerged during the nineteenth century regarding public health?
11. How is our perception of war influenced by the mass media? Give examples.
12. Who was Rosie the Riveter?
13. How did the American public receive news about Vietnam?
14. Why were images of September 11, 2001 such a powerful force?
15. Who controls public opinion?
16. What effect do large conglomerates have on communications media?
17. Who are the gatekeepers of public opinion?
18. What influence has reality television had on the viewing public?

RELATED ACTIVITIES

1. Survey your friends and relatives, asking them where they get the news that is important to them. Do they rely solely on interpersonal communications, or do they use the newspapers, television, and radio? How frequently do they use these various media during a given week? Report your findings.

2. Think about one television show with violent, aggressive behavior and one without it. Write a brief narrative describing your feelings after each show you watched.

3. Make a list of your family members and closest friends. Based on what you know, next to each name, write the kind of music that person likes to listen to. What does this tell you about the influence radio and music have on people's lives?

---

4. What kind of music do you most enjoy? How does the mass media portray people who listen to this type of music? Do you "fit" the portrayal?

5. Identify and list two examples of each of the dimensions of technology.

6. Identify at least two technological innovations in your home, then describe two positive and two negative effects of each. You may include physical, social, and psychological effects.

PRACTICE TEST

Multiple Choice
Choose the BEST answer.

1. By the time of World War II, many people formed their visual perceptions about war from _____.
   A. television
   B. newsreel footage shown in movie theaters
   C. journalists
   D. all of the above

2. The media coverage of World War II was NOT primarily oriented towards _____.
   A. portraying the United States as a world hero
   B. portraying the enemy as evil
   C. generating support for the war effort
   D. showing the public what war was really like

3. _____ wore a bandanna around her head and was sweating as she worked in a factory supporting the war effort during World War II.
   A. Rosie the Riveter
   B. Rosanna the Riveter
   C. Susie the Supporter
   D. Faith the Freedom Worker

4. Which conflict was really the first one that people saw a realistic portrayal of killing and fighting?
   A. Vietnam
   B. World War II
   C. Korean War
   D. Gulf War

5. Seeing the real images of war and conflict played a big role in creating _____ about the Vietnam conflict.
   A. support among Americans
   B. division among Americans
   C. loyalty among Americans
   D. all of the above

6. One of the media-related events that made September 11, 2001, such a powerful force in the world is _____.
   A. a variety of sources, video and cell phones, provided us with a sense of what happened very quickly after it happened
   B. images were controlled by the government
   C. it took place on American soil
   D. many nationalities were represented among those who lost their lives

7. Which of the following is true about the social institution of communications media in America?
   A. It informs us.
   B. It instructs us.
   C. It entertains us.
   D. All of the above.

8. Which of the following is an accurate statement about the social institution of communications media?
   A. Ownership is increasing in diversity.
   B. More independent companies are being formed.
   C. It is controlled by fewer and fewer companies.
   D. None of the above.

9. Often, media conglomerates are considered _____ of what we're allowed to see and hear.
   A. generators
   B. gatekeepers
   C. controllers
   D. monopolies

10. A hybrid form of television entertainment has bridged the gap between fact and fiction. This is called _____.
   A. reality television
   B. enterfiction television
   C. entertainment television
   D. spoof television

11. One of the issues with some reality television is that it often displays _____.
   A. personal circumstances
   B. bias toward certain groups
   C. perverts and deviants
   D. a weird sense of the truth

12. Technological _____ is used to refer to the fact that technological changes often have both positive and negative effects.
   A. paradox
   B. dualism
   C. effectism
   D. none of the above

13. According to Judith Perrolle in the video lesson, which of the following is a promise that often comes with new technologies?
   A. Making things better
   B. Making things more convenient
   C. Making things more equitable
   D. All of the above

14. Which of the following technologies could create technological dualism in our lives?
    A. Computers
    B. Answering machines
    C. Cell phones
    D. All of the above

15. According to Judith Perrolle in the video lesson, an important aspect of any new technology is _____.
    A. convenience
    B. the way it is used
    C. cost
    D. features

16. Which of the following is NOT a dimension of technology?
    A. Technological tools, instruments, and gadgets
    B. Body of technical skills
    C. Organizational networks
    D. Hierarchies and careers

17. Which of the following is a catalyst for major social change?
    A. Population growth
    B. Technological innovation
    C. Epidemics
    D. All of the above

18. Technological _____ refers to efforts to anticipate the consequences of particular technologies for individuals and for society as a whole.
    A. dualism
    B. theory
    C. change
    D. assessment

19. Some sociologists think a new technological boom is occurring based on _____.
    A. computers
    B. telecommunications
    C. biotechnologies
    D. all of the above

20. _____ refers to the effects of society on the natural environment.
   A. Pollution
   B. Technological stress
   C. Environmental stress
   D. None of the above

Essay Questions
Answer the following questions using complete sentences in a well-developed essay.

1. Discuss the dimensions of technological dualism.

2. Discuss an example of environmental stress.

ANSWER KEY

The following provides the answers and references for the practice test questions.

Multiple Choice:
1. B..............LO 1 .... video segment 2
2. D..............LO 1 .... video segment 2
3. A..............LO 1 .... video segment 2
4. A..............LO 1 .... video segment 2
5. B..............LO 1 .... video segment 2
6. A..............LO 1 .... video segment 2
7. D..............LO 2 .... student course guide lesson overview, p. 183; video segment 4
8. C..............LO 2 .... video segment 4
9. B..............LO 2 .... video segment 4
10. A..............LO 2 .... video segment 5
11. B..............LO 2 .... video segment 5
12. B..............LO 3 ....textbook, pp. 533–535; video segment 6
13. D..............LO 3 .... video segment 6
14. D..............LO 3 .... video segment 6
15. B..............LO 3 .... video segment 6
16. D..............LO 3 ....textbook, p. 533
17. D..............LO 3 ....textbook, pp. 535–537
18. D..............LO 3 ....textbook, p. 536
19. D..............LO 3 ....textbook, pp. 536–537
20. C..............LO 3 ....textbook, p. 538; video segment 6

Essay Questions

Answers should include the following types of statements/points:

1. ....LO 3..textbook, pp. 533–535
   - Technological tools, instruments, machines, and gadgets which are used in accomplishing a variety of tasks. These material objects are best referred to as apparatus, the physical devices of technical performance.
   - The body of technical skills, procedures, routines—all activities or behaviors that employ a purposive, step-by-step, rational method of doing things
   - The organizational networks associated with activities and apparatus

2. ....LO 4...textbook, p. 538; video segment 6
   Snowmobile use in frozen lakes compacts the snow. This reduces amount of sunlight filtering through to maintain plant life underwater. Decomposition of plant life robs water of oxygen. The fish then cannot survive, but die from asphyxiation.

LESSON CONTRIBUTORS

Steve Blow, Journalist, Dallas Morning News, Dallas, TX

Raymond A. Eve, Professor of Sociology, University of Texas—Arlington, Arlington, TX

Gordon Fellman, Professor of Sociology, Brandeis University, Waltham, MA

Judith A. Perrolle, Associate Professor, Northeastern University, Boston, MA

Bob Ray Sanders, Vice President and Associate Editor, Fort Worth Star-Telegram, Fort Worth, TX

Timothy L. Sullivan, Professor of Sociology and Anthropology, Cedar Valley College, Lancaster, TX

Brian Uzzi, Professor of Sociology, Northwestern University, Evanston, IL

Mary Virnoche, Assistant Professor, Humboldt State University, Denver, CO

Lesson 20

Population and Urbanization

OVERVIEW

Around the world, more than four children are born every second. In about an hour's time, 16,122 babies will become members of the human family. While you sleep, 128,976 babies will come into this world. In a day's time, the world's population will increase by 387,000 (even though thousands will also die).

Sociologists study population change and population growth in an effort to help communities and urban areas plan for the future. They also study characteristics of populations and how they are distributed. There are a variety of factors that affect population growth including birth rate, death rate, and even the status of women in the workforce. When women enter the workforce and delay childbearing, it can greatly affect the birth rate of a population.

When sociologists study the characteristics of a population, they look at demographic factors to determine the age, ethnic, and gender diversity of the population. These characteristics are critical to adequate planning for the future. Is the population primarily one that represents young adults who are having children? If so, what are the implications for future educational needs in the community? Will more children mean that more schools need to be built to accommodate the educational needs of those children?

As populations expand and urban areas become *home* to many in the population, how an urban area develops is important to sociologists. Sociologists have developed urban growth models to help explain the patterns of expansion that apply to urban areas. One of the earliest models was the *concentric zone model*, which developed in Chicago. But the concentric zone model has become outdated because of urban expansion.

Today, sociologists frequently help city planners apply the *peripheral model of urban expansion* to cities. The best way to understand the peripheral model is to apply it to an actual city. In the lesson video, you will learn how this model is reflected in the growth patterns and urban expansion of Houston, Texas, as Houston city planner Jerry Wood talks about his city.

Suburbs grew out of urban expansion as people sought a new lifestyle, away from the crowded, downtown urban centers. What factors contributed to the growth of suburbs in America? In the lesson video, you will hear sociologists talk about why suburban living has become so popular.

In many communities today, there is a constant struggle between the needs of the people and the needs of the urban landscape. Environmental concerns—pollution, congestion, lack of open, green spaces, as well as issues related to a shortage of housing and segregated neighborhoods—are some of the problems that plague many of our urban landscapes. These are the issues that compel sociologists to explore population change and urban expansion.

LESSON ASSIGNMENTS

Text: William Kornblum: *Sociology in a Changing World*, Chapter 21, "Population, Urbanization, and the Environment" pp. 551–581

Video: "Population and Urbanization" from the series *Exploring Society, Introduction to Sociology*

LESSON GOAL

After completing this lesson, you will know why sociologists study the relationship of population to urbanization.

LESSON LEARNING OBJECTIVES

1. Explain why the study of population is important.
2. Using an example, discuss models of urban expansion.
3. Discuss the factors that influenced urban growth.
4. Discuss issues related to urbanization.

REVIEW

The following steps are intended to help you learn the material in this lesson. To maximize your learning experience:

a. Scan the Lesson Focus Point questions.
b. Read the assigned text pages.
c. View the video.
d. Write answers to the Lesson Focus Point questions.
e. Complete the Related Activities assigned by your instructor. If none are assigned, use them to help you review the lesson material.
f. Take the Practice Test and check your answers with the Answer Key located at the end of the lesson.

LESSON FOCUS POINTS

1. What does the term *population explosion* mean?
2. How do sociologists measure and predict population changes?
3. Why are sociologists interested in population and urbanization?
4. What are some of the things that contribute to population growth and urbanization?
5. What problems are associated with urbanization?
6. What is the *crude birth rate*? *Crude death rate*? *Rate of reproductive change*?
7. What is *demographic transition*?
8. What are the stages of demographic transition? Primary features of each stage?
9. What is the population of the United States? What are the projections?
10. What are some environmental considerations that come with population growth?
11. What contributed to suburban growth?
12. What are *metropolitan areas*?
13. What is the most populated urban city?
14. What is the *concentric zone theory*? *Peripheral model*?
15. What has caused the development of suburban areas?
16. What are *satellite cities*?
17. What is *strip development* and *sprawl*?
18. What is a *megalopolis*?
19. What is *decentralization*?
20. What are environmental concerns that come with urbanization?
21. What are *private communities*?
22. What are *defended neighborhoods*?
23. What is *gentrification*?

RELATED ACTIVITIES

1. Find someone in your family or a friend's family who remembers life "back on the farm." Ask the person to describe the way it was—the dependence on subsistence farming, use of animal power, importance of the village, and so forth. Summarize the results of your investigation.

2. Describe situations or areas in the United States that meet the criteria of a peripheral region. What are the conditions that tend to make it a peripheral region? Describe them graphically.

3. What is your community doing to deal with its waste? Are there problems in coping with increasing volumes of garbage? Are there efforts to recycle? Record the results of your inquiry.

PRACTICE TEST

Multiple Choice
Choose the BEST answer.

1. Populations change as a consequence of all of the following, EXCEPT _____.
   A. births
   B. deaths
   C. migration
   D. environment

2. Which of the following is NOT a factor that demographers use to explore population change?
   A. Crude birth rate
   B. Crude death rate
   C. Birth control
   D. Migration

3. _____ is a set of major changes in birth and death rates that has occurred most completely in urban industrial nations.
   A. Demographic transition
   B. Demographic tangible
   C. Population growth
   D. Population rate

4. No population has entered the third stage of demographic transition without limiting its _____ rate.
   A. death
   B. growth
   C. birth
   D. all of the above

5. Which of the following is NOT one of the stages of demographic transition?
   A. High growth potential stage
   B. Transitional growth stage
   C. Declining growth stage
   D. Incipient decline stage

6. _____ refers to the proportion of the total population that is concentrated in urban settlements.
   A. Centralization
   B. Central city zone
   C. Urbanization
   D. Revitalization

7. One of the primary reasons sociologists are interested in studying population is because research _____.
   A. shows the impact of population growth
   B. reveals stresses on earth's natural resources
   C. reveals the rapid growth of cities
   D. all of the above

8. The distribution of population and the changes that come with population growth might affect _____.
   A. political representation
   B. the wealth of the population
   C. the ethnicity of the population
   D. the gender roles of the population

9. Which of the following factors may be responsible for pushing people out of rural areas?
   A. Lack of opportunities to obtain farmland
   B. The seasonal nature of employment in agriculture
   C. The pull of family living in urban areas
   D. All of the above

10. Why do sociologists use models of urban expansion?
    A. To understand urban growth
    B. To help predict modern urban growth
    C. To help plan for the future
    D. All of the above

11. Which of the following is NOT a model used to examine urban growth?
    A. Concentric zone
    B. Satellite cities growth
    C. Strip development and sprawl
    D. Intersections theory

12. The _____ model was originally developed to understand urban expansion in _____.
    A. Concentric zone; Philadelphia
    B. Concentric zone; Chicago
    C. Strip model; Chicago
    D. Peripheral model: Chicago

13. _____ involves the growth of metropolitan areas in which outlying areas become more important at the expense of the central city.
    A. Decentralization
    B. Suburban growth
    C. Inter-city sprawl
    D. Reverse centralization

14. In the peripheral model, what are the two types of highway development?
    A. Intrastate; interstate
    B. City; county
    C. Toll; unrestricted
    D. Radial; circumferential

15. _____ usually include government facilities such as post offices, tax offices, and fire department.
    A. Open spaces
    B. Green belts
    C. Shopping malls
    D. Service centers

16. In the lesson video, according to sociologist, Mary Pattillo the rise of the suburbs was directly related to a housing crises and _____.
    A. the promotion of VA Loans by the federal government
    B. returning servicemen and servicewomen
    C. immigration
    D. all of the above

17. With the initial rise of the suburbs, many feared the _____.
    A. rise of crime
    B. congested living spaces
    C. cookie-cutter sameness
    D. all of the above

18. Which of the following problems are influenced by urbanization?
    A. Housing and educating the people moving into the urban areas
    B. Caring for the health of people moving into the urban areas
    C. Preventing gang violence and intergroup hatred
    D. All of the above

19. The process where poor neighborhoods are renovated by higher-income newcomers while poor residents and merchants are pushed out, is known as _____.
    A. gentrification
    B. defender neighborhood
    C. invasion
    D. neighborhood stratification

20. The most common source of intergroup conflict in American cities has been _____.
    A. gentrification
    B. invasion
    C. racial tension
    D. defended neighborhoods

Essay Questions

Answer the following questions using complete sentences in a well-developed essay.

1. Describe at least two examples of ways in which residents of your neighborhood or other neighborhoods in your city try to "defend" the neighborhood from "invasion."

ANSWER KEY

The following provides the answers and references for the practice test questions.

Multiple Choice:

1.	D	LO 1	textbook, p. 553
2.	C	LO 1	textbook, p. 553
3.	A	LO 1	textbook, p. 553
4.	C	LO 1	textbook, p. 554; video segment 3
5.	C	LO 1	textbook, pp. 553–556
6.	C	LO 1	textbook, p. 559
7.	D	LO 1	textbook, p. 552
8.	A	LO 1	video segment 2
9.	D	LO 2	textbook, p. 561
10.	D	LO 2	textbook, p. 565; video segment 3
11.	D	LO 2	textbook, pp. 565–568
12.	B	LO 2	textbook, pp. 565–567; video segment 3
13.	A	LO 2	textbook, p. 569–570
14.	D	LO 2	video segment 3
15.	D	LO 2	video segment 3
16.	A	LO 3	video segment 5
17.	C	LO 3	video segment 5
18.	D	LO 4	textbook, p. 563
19.	A	LO 4	textbook, p. 575
20.	C	LO 4	textbook, p. 576

Essay Questions

Answers should include the following types of statements/points:

1. ....LO 4 ....textbook, p. 576
   - Zoning (regulations which establish minimum lot sizes)
   - Neighborhood improvement groups which might be in the form of vigilante action or street corner gangs
   - Cost of homes
   - Regulations residents must conform (agree) to within an area
   - "Closed," private neighborhoods

LESSON CONTRIBUTORS

Paul N. Geisel, Professor of Urban Affairs, University of Texas—Arlington, Arlington, TX

Mary Pattillo, Associate Professor of Sociology, Northwestern University, Evanston, IL

Terry Williams, Associate Professor, New School University, New York, NY

Jerry Wood, Deputy Assistant Director, Planning and Development Department, City of Houston, Houston, TX

Bette Woody, Professor of Sociology, University of Massachusetts, Boston, MA

Alford Young, Jr., Assistant Professor of Sociology, University of Michigan, Ann Arbor, MI

Special acknowledgement and thanks go to:

Greater Houston Convention and Visitors Bureau, Houston, TX

Lesson 21

Social Change

OVERVIEW

Societies are continuously exposed to social change. Such change occurs in our social institutions, our populations, our social structures, and our culture. Whether from internal societal forces or external forces, social change is inevitable and affects our lives. It is the sociologist who explores social change because of its powerful influence in our lives.

You will meet sociologists who define social change and talk about the three levels of sociological analysis. These levels of analysis can be understood when examining how computer technology has changed the ways in which society communicates with others, works, and delivers educational opportunities.

War continues to be one of the most powerful forces of social change. It can topple governments, destroy social institutions, and alter entire cultures. But war is a paradox—it is both a destructive *and* creative force. Whether in its effect on our economy or in the roles and statuses of women and ethnically diverse populations, World War II brought far-reaching social change to the United States, and its impact continues to be felt today.

Modernization is a process that affects societies at varying rates. One area of society that has been greatly influenced by modernization is agriculture. In the video lesson, you will meet Jerry Hare, a third-generation farmer who has lived through the many changes that have come with modernization. How has modernization affected the small farmer? What are the positive aspects that modernization has brought? What about the negative aspects?

Technology, war, and modernization—all are powerful social forces that continue to bring social change in our society. Our lives are not the lives of our parents, and our children's lives will be different than ours. Such is the influence of social change.

LESSON ASSIGNMENTS

Text: William Kornblum: *Sociology in a Changing World*, Chapter 22, "Global Social Change," pp. 583-610

Video: "Social Change" from the series *Exploring Society, Introduction to Sociology*

LESSON GOAL

After completing this lesson, you will be able to discuss the dynamics of social change.

LESSON LEARNING OBJECTIVES

1. Discuss the effects of social change from the three levels of sociological analysis.
2. Discuss war as a powerful force of social change.
3. Discuss George Ritzer's *Cathedrals of Consumption*.
4. Discuss modernization as a powerful force of social change.
5. Explain how significant social changes in society affect "everyday" life.
6. Discuss the models of large-scale social change.

REVIEW

The following steps are intended to help you learn the material in this lesson. To maximize your learning experience:

a. Scan the Lesson Focus Point questions.
b. Read the assigned text pages.
c. View the video.
d. Write answers to the Lesson Focus Point questions.
e. Complete the Related Activities assigned by your instructor. If none are assigned, use them to help you review the lesson material.
f. Take the Practice Test and check your answers with the Answer Key located at the end of the lesson.

LESSON FOCUS POINTS

1. What is the definition of *social change*?
2. How do internal (endogenous) forces differ from external (exogamous) forces of social change?
3. Population growth occurs at what level of social change?
4. The changing norms of behavior occur at what level of social change?
5. Urbanization is an example of what level of social change?
6. At what level of social change are community boundaries blurred?
7. Is social change universal?
8. Are social change and progress synonymous?
9. Can social change be predicted or controlled?
10. What are the two forces of social change?
11. What are the ecological effects of war on human populations?
12. How does war change a society's culture and social institutions?
13. What is *modernization*?
14. According to Neil Smelser in the textbook, what are the changes associated with modernization?
15. What is *cultural lag*?
16. What is *postmodernism*?
17. What is a developing nation? Modernizing nation?
18. What are indicators of social change?
19. According to the textbook, what is Immanuel Wallerstein's world system theory?
20. What are the components of the evolutionary model of social change?
21. What are cyclical theories of social change?
22. What is the conflict model of social change?
23. What is the functionalist model of social change?
24. Describe the effects of social change.
25. Identify the models of social change.

RELATED ACTIVITIES

1. Outline a social problem that has been caused, at least partially, by changes in science and technology.

2. Discuss three technological changes that have impacted your life. Describe your *initial* reaction to each change. Explain the differences each technological change has made in your life—both positive and negative. (Example: Initially, I was frightened of computers. Now, I can produce papers with images, tables, title sheets, tables of contents, and spelling errors corrected by the computer. Unfortunately, I find myself spending more time on my computer and less time with my family.)

PRACTICE TEST

Multiple Choice
Choose the BEST answer.

1. According to your textbook, which of the following is NOT an aspect of social change?
   A. The ecological ordering of populations and communities.
   B. The patterns of roles and social interactions.
   C. The structure and functioning of institutions and culture.
   D. None of the above.

2. _____ social forces are those that come from outside a society.
   A. Endogenous
   B. Exogenous
   C. Externalized
   D. Internalized

3. Which of the following statements about social change does NOT belong?
   A. Social change is a positive development for a society.
   B. Social change can be very difficult for a society.
   C. Social change can be very disruptive for a society.
   D. Social change is inevitable.

4. Changes in the social structures of a society would be an example of
    _____ social change.
    A. micro
    B. macro
    C. middle
    D. none of the above

5. Being able to send an e-mail to family members is an example of how the
    computer has brought about this _____ level of social change.
    A. macro
    B. middle
    C. micro
    D. personal

6. Two of the most powerful forces of social change are _____ and _____.
    A. technology; science
    B. technology; war
    C. war; modernization
    D. modernization; technology

7. Which of the following is an example of the cultural impact of war?
    A. Soldiers who have been maimed returning home
    B. Post-traumatic stress disorder
    C. Survivor guilt
    D. All of the above

8. For African Americans and other nonwhites, World War II _____.
    A. showed little change in status/prestige
    B. greatly changed roles and statuses
    C. did not change the way people perceived those groups
    D. none of the above

9. World War II sparked the government's role of _____.
    A. assuming more responsibility for society
    B. diminishing jobs for women
    C. decreasing jobs for nonwhite citizens
    D. all of the above

10. Cathedrals of consumption have the purpose of _____.
    A. reinforcing religion
    B. providing a vacation
    C. allowing families to spend quality time together
    D. making you spend your money

11. According to the video lesson, George Ritzer used the term cathedrals of consumption because of _____.
    A. a shared magical quality they have with churches
    B. the sense that both religious and consumption cathedrals are worshipped in our society
    C. both religious and consumption cathedrals are often spectacular
    D. all of the above

12. Modernization refers to all the changes that societies and individuals experience as a result of _____, urbanization, and the development of nation-states.
    A. technology
    B. science
    C. status roles
    D. industrialization

13. Technology provided society with the use of machines in industry, which greatly increased worker output. Later, specific guidelines and safeguards for the workers using those machines were developed. This is an example of

    _____.
    A. innovation
    B. modernization
    C. cultural lag
    D. technological dualism

14. According to the video lesson, Jerry Hare and his family have experienced modernization in _____.
    A. urbanization
    B. agriculture
    C. business
    D. family life

15. Modernization has brought the use of _____ to agriculture.
    A. chemical fertilizers
    B. pesticides
    C. herbicides
    D. all the above

16. Which of the following is NOT a significant social change in Western societies?
    A. Gender roles
    B. Racial/ethnic relations
    C. Public policy
    D. None of the above

17. Which of the following accounts for an explanation for the growing divergence between the haves and the have-nots?
    A. Changes in the economic structure of society.
    B. Changes in ethnic relations.
    C. Changes in gender roles.
    D. None of the above.

18. _____ are laws and regulations that are formulated by governments to control, regulate, or guide behavior.
    A. Statutes
    B. Operational points
    C. Public policies
    D. Governmental policies

19. Which of the following is NOT a main component of the evolutionary model of social change?
    A. Social change is natural and constant.
    B. Social evolution has a direction.
    C. Social evolution is continuous.
    D. Change is unnecessary.

20. _____ models of social change argue that conflict among groups with different amounts of _____ produces social change.
    A. Conflict; wealth
    B. Evolutionary; power
    C. Cyclical; power
    D. Conflict; power

Essay Question
Answer the following question using complete sentences in a well-developed essay.

1. Discuss the possible impact of war on a society.

ANSWER KEY

The following provides the answers and references for the practice test questions.

Multiple Choice:
1. D.............LO 1 ....textbook, p. 584
2. B.............LO 1 ....textbook, p. 584
3. A.............LO 1 ....textbook, pp. 584–587
4. B.............LO 1 .... video segment 2
5. C.............LO 1 .... video segment 2
6. C.............LO 2 ....textbook, p. 587
7. D.............LO 2 ....textbook, pp. 589–590
8. B.............LO 2 .... video segment 3
9. A.............LO 2 .... video segment 3
10. D.............LO 3 .... video segment 4
11. D.............LO 3 .... video segment 4
12. D.............LO 4 ....textbook, pp. 594–595
13. C.............LO 4 ....textbook, p. 595
14. B.............LO 4 .... video segment 5
15. D.............LO 4 .... video segment 5
16. D.............LO 5 ....textbook, pp. 599–603
17. A.............LO 5 ....textbook, p. 602
18. C.............LO 5 ....textbook, p. 602
19. D.............LO 6 ....textbook, pp. 603–604
20. D.............LO 6 ....textbook, p. 605

Essay Question

Answers should include the following types of statements/points:

1. ....LO 2...textbook, p. 587–593; video segment 3

   <u>Answer:</u>
   - Population unbalance due to loss of life
   - Women become widows
   - Children left fatherless
   - Labor shortages
   - Mobilization for war
   - Death expulsion, banishment
   - Post-traumatic stress disorder
   - Changing gender roles, expectations
   - National pride, patriotism
   - Shame, guilt
   - Social institutions

LESSON CONTRIBUTORS

Raymond A. Eve, Professor of Sociology, University of Texas—Arlington, Arlington, TX

Tom Mayer, Professor of Sogiology, University of Colorado—Boulder, Boulder, CO

Mary Pattillo, Associate Professor of Sociology, Northwestern University, Evanston, IL

Judith A. Perrolle, Associate Professor, Northeastern University, Boston, MA

George Ritzer, Professor, University of Maryland,College Park, MD

Mary Virnoche, Assistant Professor, Humboldt State University, Denver, CO

Bette Woody, Professor of Sociology, University of Massachusetts, Boston, MA

Special acknowledgement and thanks go to:

Karen Ellis, Assistant Director for Public Services, Nicholson Memorial Library System, Garland, TX

Jerry Hare, Tahlequah, OK

George Susat, Waxahachie, TX

Lesson 22

Social Action

OVERVIEW

Modern societies host a multitude of conflicting interests. Business owners are interested in profits and workers want living wages. Parents are interested in protecting and caring for their children, and children are interested in having fun. Some politicians represent the interests of business enterprises; some favor protection of consumers and wage earners.

Sometimes social institutions fail to meet peoples' needs. When this happens, some people become frustrated and seek solutions through social action. Social action is a personal response or commitment that is expressed individually or collectively i.e., at small group, community, societal, and/or global levels. At the one-on-one level, an employee confronts a supervisor. At the community level, a neighborhood organizes a petition effort. At the national level, a union goes on strike. At the global level, environmental groups meet to encourage sustainable development.

People become involved in social movements to either bring about or resist changes. People in movements have a feeling of solidarity around their interests and beliefs. Mobilization of people and resources is a conscious and on-going effort in movements. Social movements always involve conflict. Social activists regularly agitate people to change. These efforts are necessary to motivate and to help those with power and authority to see the needs and then to effect changes.

In this video lesson, you will learn that social action often involves collective behavior such as rallies, demonstrations, and marches. Some collective behavior may be organized, as in a strike; however, it might involve highly emotional, spontaneous actions in unstructured situations like riots. Sociologists have a long history of studying social action, social movements, and collective behavior. It is one of the most fascinating areas of study in sociology.

You will also learn about sociologist Victor Ayala's personal journey to become one of his community's leaders in the AIDS awareness movement. His story exemplifies how the social actions of an individual can make a difference and bring about social change.

LESSON ASSIGNMENTS

Text: William Kornblum: *Sociology in a Changing World*, Chapter 9, "Collective Behavior, Social Movements, and Mass Publics," pp. 210–231

Video: "Social Action" from the series *Exploring Society, Introduction to Sociology*

LESSON GOAL

After completing this lesson, you will be able to explain how social action is related to collective behavior and social change.

LESSON LEARNING OBJECTIVES

1. Using Lofland's Typology of Spontaneous Collective Behavior, classify types of collective behavior observed at various social events.
2. Describe and give examples of the types of social movements.
3. Discuss the theories associated with collective behavior.
4. Appreciate the role of the individual in the process and the consequences of social action.

REVIEW

The following steps are intended to help you learn the material in this lesson. To maximize your learning experience:

  a. Scan the Lesson Focus Point questions.
  b. Read the assigned text pages.
  c. View the video.
  d. Write answers to the Lesson Focus Point questions.
  e. Complete the Related Activities assigned by your instructor. If none are assigned, use them to help you review the lesson material.
  f. Take the Practice Test and check your answers with the Answer Key located at the end of the lesson.

LESSON FOCUS POINTS

1. What is *collective behavior*?
2. Why are sociologists interested in studying collective behavior?
3. Identify the types of spontaneous collective behavior.
4. What is a *crowd*?
5. What is a *mass*?
6. What emotions motivate crowds and masses?
7. What is Lofland's Typology of Spontaneous Collective Behaviors?
8. What are examples of spontaneous collective behavior exemplified by Lofland's typology?
9. What are examples of crowd behavior? Mass behavior?
10. What is a social movement?
11. What are the types of social movements?
12. How do the goals of social movements differ?
13. How do the goals of a social movement determine how the social movement is classified?
14. What are examples of revolutionary movements, reformist movements, conservative movements, reactionary movements?
15. What are expressive social movements? Examples?
16. What is a *messianic* or *millenarian social movement*? Give examples.
17. What is the difference between political and social revolutions?
18. What is the *relative deprivation theory*?
19. What is a *wave of protest*?
20. According to William Gamson, what are the components of the action frames?
21. What is *charisma*?
22. What is *co-optation*?
23. How is social action related to social change?
24. How did sociologist Victor Ayala become involved in social action?

RELATED ACTIVITIES

1. List five specific non-textbook examples of collective behavior, beginning with the least organized and moving to the most organized.

2. Identify a recent or current social movement with a leader who has charisma—as defined by Max Weber. Identify a local or national leader, and describe the special qualities of that leader. Then describe situations in which the leader

---

Lesson 22—Social Action
215

appears to be inspiring followers. What does the leader say and do? What do the followers do that indicate that they are being motivated?

3. Describe a recent example of mass public behavior. In your description, give some indication of the size of the population involved. What actions, events, and situations surrounded the collective behavior?

4. Identify a current or past radical social movement, and describe the changes in the social system it is trying or tried to bring about.

PRACTICE TEST

Multiple Choice
Choose the BEST answer.

1. Which of the following is NOT an example of collective behavior?
    A. Riots in Los Angeles after the trial of the police officers accused of beating Rodney King
    B. The strikes and picketing that spurred the labor movement
    C. Demonstrations that enlivened the women's movement
    D. A television program broadcast worldwide

2. The term _____ refers to a continuum of unusual or nonroutine behaviors that are engaged in by large numbers of people.
    A. social movement
    B. collectivity
    C. fad and fashions
    D. collective behavior

3. One of the reasons sociologists are so interested in studying collective behavior is because it provides insight into _____.
    A. social change
    B. panic crowd behavior
    C. mob control
    D. all of the above

4. Which of the following is NOT a characteristic of collective behavior?
   A. It involves unusual behavior.
   B. It involves non-routine behavior.
   C. It involves behaviors that occur everyday.
   D. It involves behavior that is engaged by large numbers of people.

5. A _____ is a large number of people who are gathered together in close proximity to one another; a _____ is more diffuse and does not occur in a physical setting.
   A. mass; crowd
   B. public; crowd
   C. crowd; public
   D. crowd; mass

6. A large number of people watching the same television program, at the same time, in different locations throughout the United States is an example of a _____.
   A. mass
   B. crowd
   C. public
   D. collectivity

7. Which of the following is an example of a crowd being motivated by joy?
   A. Attendees at a Mardi Gras parade
   B. Audience at a rock concert
   C. Fans gathered in a soccer stadium for the World Cup
   D. All of the above

8. _____ are intentional efforts by groups in a society to create new institutions or reform existing ones.
   A. Social movements
   B. Fads and fashions
   C. Collectivities
   D. Spontaneous behaviors

9. The Ku Klux Klan is an example of a _____ social movement.
   A. revolutionary
   B. reformist
   C. reactionary
   D. expressive

10. Men practicing collective drumming to bond together and express suppressed emotions is an example of a(n) _____ social movement.
    A. millenarian
    B. messianic
    C. reformist
    D. expressive

11. The Civil Rights Movement is an example of a _____ social movement.
    A. reformist
    B. revolutionary
    C. reactionary
    D. conservative

12. Which of the following is an example of a social movement?
    A. Crowd behavior at a football game
    B. Mass behavior panicking
    C. Mass suicide
    D. Civil rights

13. The Million-Man March in Washington, DC, in the fall of 1995, had elements of both the _____ and a _____ social movement.
    A. reformist; reactionary
    B. reactionary; conservative
    C. reformist; expressive
    D. reactionary; expressive

14. Sociologists examine social movements by looking at _____.
    A. the nature of the change the movement seeks
    B. the amount of change the movement seeks
    C. the goals of its members
    D. all of the above

15. The American Revolution was a _____.
    A. social revolution
    B. political revolution
    C. civil revolution
    D. revolution

16. Which theory attempts to explain why a person joins a revolutionary movement by looking at the person's poverty in relation to others?
    A. Relative deprivation
    B. Poverty relatedness
    C. Deprivation ratio
    D. None of the above

17. Which of the following leaders would be an example of a charismatic leader?
    A. Mahatma Gandhi
    B. Martin Luther King, Jr.
    C. John F. Kennedy
    D. All of the above

18. What event spurred Victor Ayala's journey to better understand HIV?
    A. His child was diagnosed HIV positive.
    B. A close friend of his was diagnosed with AIDS.
    C. A family member contracted HIV.
    D. All of the above.

19. Sociologist Victor Ayala pursued his doctorate in sociology, writing his dissertation on _____.
    A. homelessness and children
    B. AIDS and homelessness
    C. HIV and gays
    D. women and AIDS

20. Which of the following statements is NOT true of Victor Ayala's work?
    A. His work began as volunteer work.
    B. He taught at a community college.
    C. He worked with AIDS patients at night.
    D. He had no time for speaking engagements.

Essay Question

Answer the following question using complete sentences in a well-developed essay.

1. Identify a current or past social movement, and describe the changes in the social system it is trying or tried to bring about.

ANSWER KEY

The following provides the answers and references for the practice test questions.

Multiple Choice:

1. A.............LO 1 ....textbook, p. 212
2. D.............LO 1 ....textbook, p. 212, video segment 2
3. A.............LO 1 ....textbook, p. 212; video segment 2
4. C.............LO 1 ....textbook, p. 212; video segment 2
5. D.............LO 1 ....textbook, pp. 214–215; video segment 2
6. A.............LO 1 ....textbook, pp. 214–215; video segment 2
7. D.............LO 1 ....textbook, p. 215; video segment 2
8. A.............LO 2 ....textbook, p. 212 ; video segment 4
9. C.............LO 2 ....textbook, p. 216; video segment 4
10. D.............LO 2 ....textbook, pp. 216–217; video segment 4
11. A.............LO 2 ....textbook, p. 216; video segment 4
12. D.............LO 2 ....textbook, p. 212 and p. 217; video segment 4
13. C.............LO 2 .... video segment 4
14. D.............LO 2 .... video segment 4
15. B.............LO 3 ....textbook, pp. 218–219
16. A.............LO 3 ....textbook, pp. 219–220
17. D.............LO 3 ....textbook, p. 222
18. C.............LO 4 .... video segment 5
19. B.............LO 4 .... video segment 5
20. D.............LO 4 .... video segment 5

Essay Question

Answers should include the following types of statements/points:

1. ....LO 2...textbook, pp. 216–220; video segment 4
   Social Movements:
   - Anti-war
   - Civil Rights
   - Environmental protection
   - Labor
   - Women's movement
   - Gained the right to vote (19th Amendment)
   - Fighting to eliminate gender-based discrimination in employment
   - Continuing pressure to end wage inequality
   - Gaining respect for "women's work"
   - Increasing access to education
   - Continuing to seek an end to sexual harassment
   - Seeking more political representation

LESSON CONTRIBUTORS

Victor A. Ayala, Professor of Sociology, New York City Technical College, Brooklyn, NY

Tracey McKenzie Elliott, Professor of Sociology, Collin County Community College District, Frisco, TX

Joshua Gamson, Associate Professor Sociology, Yale University, New Haven, CT

William Kornblum, Professor of Sociology, City University of New York, New York, NY

Parker J. Palmer, Author/Consultant, Madison, WI

Richard Scotch, Professor of Sociology, University of Texas—Dallas, Dallas, TX

Terry Williams, Associate Professor, New School University, New York, NY

Special acknowledgement and thanks go to:
  Brooklyn AIDS Task Force, Brooklyn, New York